Industrial and Commercial Heat Recovery Systems

Sydney Reiter

VNR VAN NOSTRAND REINHOLD COMPANY
NEW YORK CINCINNATI TORONTO LONDON MELBOURNE

Copyright © 1983 by Van Nostrand Reinhold Company Inc.

Library of Congress Catalog Card Number 82-17371
ISBN: 0-442-27943-4

All rights reserved. No part of this work covered by the copyright hereon may be reproduced or used in any form or by any means—graphic, electronic, or mechanical, including photocopying, recording taping, or information storage and retrieval systems—without permission of the publisher.

Manufactured in the United States of America

Published by Van Nostrand Reinhold Company Inc.
135 West 50th Street, New York, N.Y. 10020

Van Nostrand Reinhold Publishing
1410 Birchmount Road
Scarborough, Ontario MIP 2E7, Canada

Van Nostrand Reinhold
480 Latrobe Street
Melbourne, Victoria 3000, Australia

Van Nostrand Reinhold Company Limited
Molly Millars Lane
Wokingham, Berkshire, England

15 14 13 12 11 10 9 8 7 6 5 4 3 2 1

Library of Congress Cataloging in Publication Data
Reiter, Sydney.
    Industrial and commercial heat recovery systems.
    Includes index.
    1. Heat recovery.   I. Title.
TJ260.R46   1983   621.402   82-17371
ISBN 0-442-27943-4

PREFACE

Heat recovery systems are covered in a cursory manner in books on energy management. Several specialized books discuss various limited aspects of heat recovery systems. A need exists for a book on heat recovery systems sufficiently broad to cover the areas of interest to building and plant engineering and maintenance personnel, yet specific enough to provide useful application information. Industrial and Commercial Heat Recovery Systems has been written to fulfill this need.

The sections of this book have been arranged to allow their use individually, without recourse to previous chapters. The reader wishing to study a particular aspect of heat recovery systems can use the book for specific subjects. For a complete study of heat recovery systems, the reader can follow the chapter development.

Chapters 1 through 3 are introductory in nature. They are intended for the person interested in finding heat recovery applications in his building or plant, what type of system to use, and how much heat can be recovered. Chapter 4 describes various types of heat recovery equipment. Chapter 5 explains how to choose the proper size of the equipment chosen, and how to calculate heat recovery system performance. Chapter 6 provides examples of heat recovery systems to guide the reader in his choice of systems. Chapter 7 discusses various methods of financial analysis used to evaluate the system economics, and Chapter 8 contains a check list to be followed in reviewing the heat recovery system design. The sum of these chapters enables the reader to go through the design process for most heat recovery systems used in buildings and industrial plants.

iv PREFACE

All equations are derived from engineering fundamentals or explained on the basis of physical principles. The derivations are fully explained. All symbols and subscripts are listed in the beginning of the book, and design equations are tabulated at the beginning of Chapter 5 for reference in using the design methods. The reader will be able to use the design methods without any other mathematics than simple algebra.

The material presented here is the result of the author's extensive experience in the design and installation of many different types of heat recovery systems. No attempt has been made to cover specialized areas involved in process design or specialized production equipment. Rather, the book will serve as a handbook and as a reference for those interested in the practical application of heat recovery to buildings and industrial plants.

SYDNEY REITER

ACKNOWLEDGMENTS

Many individuals and firms have contributed to the preparation of this book. Heat recovery equipment photographs and data have been generously supplied by equipment manufacturers. Eric Brown, Heinz Guendner, and Norman Kolb worked with the author on many of the heat recovery systems. Greg Carroll provided advice and suggestions. Florence Parent patiently typed the manuscript and its revisions, and my wife, Connie, provided encouragement and much help. To all of them, my thanks and appreciation.

NOMENCLATURE

A	area
$ANROI$	average net return-on-investment
AP	amount purchased
B	billed cost
C	cost saving
CFI	cash flow in
CFO	cash flow out
C_p	specific heat
D	depreciation
d	density
E	effectiveness, or efficiency of existing equipment
EC	energy cost
F	percent fuel saving
H	heat recovered
h	enthalpy (heat content), or heat transfer coefficient
Δh	manometer column height
I	investment
i	percent interest
k	thermal conductivity
L	length, or life
NCF	net cash flow
NI	net income
NPB	net payback
$NROI$	net return-on-investment
OE	operating expense
PB	payback
PWF	present worth factor
P	pressure
ΔP	pressure change
ROI	return-on-investment
T	temperature

NOMENCLATURE

ΔT	change in temperature
t	time
U	overall heat transfer coefficient
u	amount of fuel energy currently used
V	volume flow
v	velocity
W	weight flow
w	moisture content
Y	year

SUBSCRIPTS

cos	change-of-state
e	exhaust, or entering
ee	exhaust entering
el	exhaust leaving
l	leaving
M	mean
std	standard
s	supply
se	supply entering
sl	supply leaving

UNITS

Btu	British thermal unit (heat to raise 1 lb of water 1°F)
cfm	cubic feet/minute
cfh	cubic feet/hour
fpm	feet/minute
fps	feet/second
ft	feet
ft^2	square feet
ft^3	cubic feet
°F	degrees Fahrenheit
gr	grains
gpm	gallons/minute
hp	horsepower
hr	hours
in.	inches
lb	pounds
min	minutes
psig	pounds/square inch gauge

CONTENTS

Preface / iii
Acknowledgments / iv
Nomenclature / v

1. **Introduction to Heat Recovery Systems / 1**
   1.1 Basic Heat Recovery / 1
   1.2 Benefits of Heat Recovery / 3
   1.3 Use of this Book / 4

2. **Amount of Heat Recovered / 5**
   2.1 Heat Recovered from or by a Gas / 6
   2.2 Heat Recovered from or by a Liquid / 6
   2.3 Heat Recovered from or by a Vapor / 7
   2.4 Heat Recovered by Pre-Heated Combustion Air / 7
   2.5 Calculation of Cost Savings / 11

3. **Heat Recovery Survey / 13**
   3.1 Location of Sources and Uses of Recovered Heat / 13
   3.2 Composition / 15
   3.3 Location / 15
   3.4 Size of Supply and Exhaust Systems / 15
   3.5 Operating Cycle / 15
   3.6 Condition of Ducts, Pipes, Equipment / 16
   3.7 Contaminants / 16

viii CONTENTS

    3.8 Installation Problems / 17
    3.9 Supply and Exhaust Operating Conditions / 17
    3.10 Heat Recovery Measurements / 18
    3.10.1 Pressure / 18
    3.10.2 Temperature / 19
    3.10.3 Flow / 21
    3.10.4 Estimating Flow Rates / 25
    3.11 Moist Air Conditions / 31
    3.12 Matching of Source and Use for Recovered Heat / 33

4. Heat Recovery Equipment / 36
    4.1 Fundamentals of Heat Exchangers / 36
    4.2 Gas/Gas Curved Plate Counter-Flow Heat Exchangers / 42
    4.3 Gas/Gas Flat Plate Counter-Flow Heat Exchangers / 46
    4.4 Gas/Gas Cross-Flow Heat Exchangers / 50
    4.5 Gas/Gas Shell-and-Tube Heat Exchangers / 54
    4.6 Gas/Gas Heat Pipe Heat Exchangers / 57
    4.7 Heat Wheels / 61
    4.8 Gas/Gas High Temperature Heat Exchangers / 66
    4.9 Combustion Air Pre-Heat Systems / 75
    4.10 Gas/Liquid Heat Exchangers / 84
    4.11 Boiler Economizers / 91
    4.12 Waste Heat Boilers / 99
    4.13 Liquid/Liquid Heat Exchangers / 107
    4.14 Air Re-circulation / 112
    4.15 Refrigeration Heat Recovery / 115
    4.16 Centrifugal Chillers and Heat Pumps / 119
    4.17 Absorption Refrigeration from Recovered Heat / 126
    4.18 Engine and Gas Turbine Heat Recovery / 132

5. Sizing and Performance of Heat Recovery Equipment / 139
    5.1 Gas/Gas Curved Plate Counter-Flow Heat Exchanger / 140
    5.2 Gas/Gas Flat Plate Counter-Flow Heat Exchanger / 143
    5.3 Gas/Gas Cross-Flow Heat Exchangers / 148
    5.4 Gas/Gas Shell-and-Tube Heat Exchanger / 153
    5.5 Gas/Gas Heat Pipe Heat Exchanger / 156
    5.6 Heat Wheels / 160
    5.7 Gas/Gas High Temperature Heat Exchanger / 165
    5.8 Combustion Air Pre-Heat / 173
    5.9 Gas/Liquid Heat Exchangers / 177
    5.10 Boiler Economizers / 189
    5.11 Waste Heat Boiler / 194
    5.12 Liquid/Liquid Heat Exchangers / 196
    5.13 Air Re-circulation / 203

6. Examples of Heat Recovery Systems / 204
    6.1   Metal Casting Plant / 204
    6.2   Roof Tile Plant / 207
    6.3   Machine Shop / 209
    6.4   Textile Dye Plant / 210
    6.5   Pet Foods Baking / 211
    6.6   Large Steel Plant / 213
    6.7   Forging Furnace / 214
    6.8   Plastics Curing Oven / 216
    6.9   Building Exhaust / 218
    6.10 Heat Recovery Systems Not Feasible / 220
    6.11 Swimming Pool Heat Recovery / 221
    6.12 Commercial Heat Pump Applications / 223

7. Economics of Heat Recovery Systems / 225
    7.1   Estimation of Heat Recovery System Cost / 225
    7.2   Payback / 226
    7.3   Return-on-Investment / 227
    7.4   Cash Flow Analysis / 227
    7.5   Net Payback / 232
    7.6   Net Return-on-Investment / 232
    7.7   Average Net Return-on-Investment / 233
    7.8   Discounted Cash Flow / 233
    7.9   Net Present Value / 234
    7.10 Internal Rate-of-Return / 234
    7.11 Example / 235

8. A Heat Recovery System Checklist / 238

Index / 241

1
INTRODUCTION TO HEAT RECOVERY SYSTEMS

Heat recovery systems are designed to conserve energy by re-using available waste heat. They transfer heat from sources of waste heat to uses for the recovered heat, with various types of heat recovery equipment. The conservation of energy not only reduces our dependence on imported fuels, but produces a cost saving to pay back the system cost. The cost saving increases as fuel costs increase, creating an inflation-proof investment.

1.1 BASIC HEAT RECOVERY

The elements of a heat recovery system are shown in Figure 1.1-1. The "source" produces waste heat as a result of a process or building operation. The waste heat can be contained in a gas, liquid, or vapor. Its temperature may be very high, as in the exhaust from a furnace, or it may be close to ambient temperature, as in the exhaust from a building ventilator. The "use" consumes heat as part of its operation. Besides liquid, gas, and vapor, the use can encompass process materials pre-heated before entering the process. The "heat recovery equipment" indicates a means for transferring the waste heat from the source in a form acceptable by the use. The type of device used for the heat recovery equipment depends upon the nature of the source and use, and their respective temperatures. Examples of heat recovery equipment include heat exchangers, waste heat boilers, and boiler economizers. Following the heat recovery equipment, the exhaust fluid is either vented or drained. The supply is received from outside sources, return from the process, or building return air. A simple example of a heat recovery system is shown in

2 INDUSTRIAL AND COMMERCIAL HEAT RECOVERY SYSTEMS

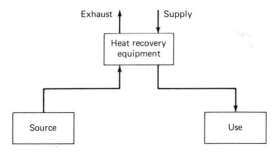

Figure 1.1-1. Elements of a heat recovery system.

Figure 1.1-2. As shown, exhaust gases from the boiler stack pass through an economizer in the stack (details of economizer construction will be found in Chapter 4). The economizer reduces the gas temperature. Returning boiler feedwater is heated in the economizer before entering the boiler. The cost saving produced by the stack economizer appears as reduced fuel to pre-heat the boiler feedwater.

The example of Figure 1.1-2 illustrates the principles of heat recovery:

1. The exhaust temperature falls as a result of losing recoverable waste heat.
2. The supply temperature rises as a result of using recovered waste heat.
3. The amount of recovered waste heat given up by the exhaust must equal the amount of recovered waste heat gained by the supply.
4. The amount of recovered waste heat is less than the total amount of exhaust heat.

Since the presence of heat in a material is measured by the material's temperature, the material's loss or gain of heat is reflected by its change in temperature. The greater the amount of heat loss or gain, the greater the

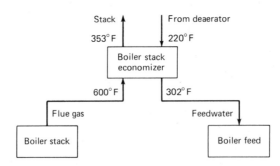

Figure 1.1-2. Boiler feedwater heat recovery system.

change in temperature. The heat loss from the exhaust must appear in the heat gained by the supply, based upon the basic laws of thermodynamics. There is no other place for the heat to exist. The temperatures of the exhaust and supply are not necessarily the same because they may originate under very different conditions, but the amount of heat lost and gained is always the same. Finally, the heat recovery equipment can recover only a portion of the total heat in the exhaust. It can reduce the exhaust temperature to the extent permitted by the supply and exhaust temperatures and by its design. The supply temperature is the lowest theoretical exhaust temperature leaving the equipment, because the supply cannot cool the exhaust below its own temperature. The equipment design fixes how closely the actual exhaust temperature approaches the supply temperature (this is more fully explained in Chapter 4). Any remaining source heat after the heat recovery equipment is lost in the exhaust leaving the heat recovery equipment.

In the example of Figure 1.1-2, the flue gases from the boiler are cooled from 600° F to 353° F in the stack economizer. The returning boiler feedwater from the process or heating system is heated in the stack economizer from 220° F to 302° F before entering the boiler. Heat remaining in the flue gas is exhausted from the stack.

1.2 BENEFITS OF HEAT RECOVERY

Heat recovery benefits fall into three categories:

1. Reduction of energy cost
2. Reduction of equipment cost and size
3. Reduction of energy use

Reduction of energy cost is the primary benefit of heat recovery. Any heat recovered from the exhaust and returned to the supply need not be supplied by purchased energy. Further, any increases in energy prices result in increased heat recovery benefits. As an example, a heat recovery installation saving $10,000 during its first year of operation will be saving $35,178 during its tenth year of operation, assuming a 15% annual energy price increase. The total ten-year saving will be $203,037.

A heat recovery system is both inflation- and price increase-proof. Very few other investments are so free of economic risk.

Additional cost benefits for heat recovery systems are available as equipment cost and size reduction. The use of recovered heat reduces the amount of heat furnished from purchased energy. Oil and gas supply pipes, electrical facilities, burners, boilers, and support structures often can be reduced. If standby facilities are required, temporary, rather than permanent,

4 INDUSTRIAL AND COMMERCIAL HEAT RECOVERY SYSTEMS

equipment can be provided, preserving the cost reduction. The reduction of heat requirements can permit greater utilization of existing process or ventilation equipment. Increased amounts of product or ventilation can be handled without increasing energy use and without new equipment. Where cyclical or peaking conditions are present, heat recovery allows a flexible way of accommodating periods of high heat demand without providing additional heating facilities.

Energy shortages have become part of the risks of operating a business or a facility. Oil embargoes, gas shortages, and electricity brown-outs can reduce production or facility operation. When energy supply is curtailed and supply allocations are established during an emergency, the amount of energy conserved by a heat recovery system is available to keep other processes or facilities in operation. Heat recovery systems are useful in avoiding expansion of service facilities. Where additional energy use might force the installation of boilers, electrical sub-stations, and so on, heat recovery systems can provide the additional energy requirement.

1.3 USE OF THIS BOOK

This book provides a systems approach to heat recovery, combined with chapters on specific design steps of heat recovery systems. Starting with an introductory chapter, the necessary formulas and methods of calculation are derived in Chapter 2. The steps of finding a suitable heat recovery system, choosing the equipment, and calculating its performance follow in Chapters 3, 4, and 5. Chapter 6 presents schematics of a wide variety of heat recovery systems, providing examples for the reader's own system design. In Chapter 7, the methods of financial analysis for economic justification are explained, completing the design process. In Chapter 8 there is a checklist of steps involved in the design of a complete heat recovery system. The chapters are self-contained to allow the reader to skip those areas not involved in his work, and to concentrate on specific subjects.

The specific information on heat recovery equipment presented in Chapter 4 is intended to be representative of each equipment category. Inclusion or omission of any manufacturer's equipment has no significance other than the availability of data suitable for publication in this book. No responsibility is implied for the accuracy of specific product information, which is solely the responsibility of the various equipment manufacturers.

Heat recovery systems and equipment included here are those intended for industrial and commercial use, either as original equipment or as a retrofit. No information is included on chemical, petrochemical, power plant, or metallurgical processes, in which the heat recovery equipment and systems are an integral part of the processes.

2
AMOUNT OF HEAT RECOVERED

The amount of heat recovered from an exhaust system and transferred to a supply system can produce two changes in their properties:

1. The heat will flow from the higher temperature to the lower temperature, reducing the temperature difference.
2. If the temperature change passes through a region of change of state, the heat required to produce the change of state will be part of the heat recovery process. For example, if water is heated and boiled, the heat consumed will be the sum of the heat used to raise the water to its boiling point, and the heat used to boil it.

The amount of heat contained by a material of weight W at a temperature T with a specific heat C_p is defined as

$$H = C_p W T. \qquad (2\text{-}1)$$

If the temperature of the material is changed by an amount ΔT, the change of its heat content is

$$H = C_p W \Delta T. \qquad (2\text{-}2)$$

When the weight, W, refers to a medium flowing at a given hourly rate in lb/hr, and ΔT is in °F, then Equation 2-2 has units of Btu/hr. If the volume

6 INDUSTRIAL AND COMMERCIAL HEAT RECOVERY SYSTEMS

flow, V, and density, d, are known, their product is the weight, W, and Equation 2-2 becomes

$$H = C_p V d \, \Delta T. \tag{2-3}$$

Should the temperature change, ΔT, pass through a region of change-of-state, the heat of condensation or vaporization per pound of medium, $h_{\cos}$, is added to Equation 2-3:

$$H = C_p V d \, \Delta T + V d h_{\cos}. \tag{2-3a}$$

2.1 HEAT RECOVERED FROM OR BY A GAS

The heat recovered from or by a flowing gas is given by Equation 2-3. The quantities are those actually present. When the volume flow is referred to standard conditions of 70° F and one atmosphere pressure, it is designated "standard cfm." For the special case of air at standard conditions, Equation 2-3 can be greatly simplified. The standard specific heat of air is 0.24 Btu/lb/°F. The standard density of air is 0.075 lb/ft^3. Substituting these quantities into Equation 2-3 and solving for heat recovered per hour:

$$H = (0.24) \, V_{\text{std}} (0.075)(60) \, \Delta T \tag{2-4}$$

$$= 1.08 \, V_{\text{std}} \, \Delta T \text{ Btu/hr.}$$

The actual volume flow is corrected to the standard volume flow by using Equation 2-5:

$$V_{\text{std}} = V \left(\frac{530}{460 + T} \right) \left(\frac{P + P_{\text{std}}}{P_{\text{std}}} \right). \tag{2-5}$$

Degrees Fahrenheit are converted to absolute temperature by adding 460 (70° F becomes 530° absolute). The value of P_{std} depends on the units used for P. If P is in psig, P_{std} is 14.7 psig. If P is in in. mercury, P_{std} is 29.92 in. mercury. Normally, this correction is unnecessary for heat recovery calculations, unless appreciable altitudes or process pressures are involved.

2.2 HEAT RECOVERED FROM OR BY A LIQUID

For most heat recovery systems involving liquid flow, the flow volume rate, V, is measured in gallons per minute, gpm. For the special case of water, the specific heat is 1 Btu/lb/°F and the specific gravity is 8.33 lb/gal. Substituting these quantities into Equation 2-3:

$$H = (1) \ V \ (8.33) \ (60) \ \Delta T$$
$$= 500 \ V \ \Delta T \ \text{Btu/hr.} \qquad (2\text{-}6)$$

2.3 HEAT RECOVERED FROM OR BY A VAPOR

The heat resulting from a change-of-state, h_{cos}, is found from published tables of vapor properties. The most common tables are the steam charts, shown in Figure 5.10-1 (Chapter 5). Entering the chart at the given pressure or temperature, the heat of vaporization is found and used in Equation 2-3a. This value of h_{cos} is the amount of heat in Btu/lb used to vaporize 1 lb of water from water at its boiling point. Conversely, the same value of h_{cos} is given up by 1 lb of steam condensing from its vapor state to its liquid state. If the steam or vapor is to be heated above its boiling temperature, or cooled from a point above its boiling temperature, then a table of superheated steam or vapor properties is used to find the amount of heat involved between the change-of-state temperature and the final temperature.

2.4 HEAT RECOVERED BY PRE-HEATED COMBUSTION AIR

Fossil fuels require oxygen in the form of air to burn and release heat. The heat of combustion is used to heat the combustion products and the contents of the equipment being heated. This could be water in a boiler, bread in an oven, or bricks in a kiln. If the combustion products leave the equipment at a high temperature, close to the temperature of the contents, then there is little available heat from the combustion process. If the combustion products leave the equipment at a low temperature, transferring most of their heat to the equipment contents, then there is a large amount of available heat from the combustion process.

When the combustion air is pre-heated before entering the equipment, more heat is made available to the combustion process. Where the available heat is small (high exhaust gas temperature), the *percent* saving of fuel is large. When the available heat is large (low exhaust gas temperature), the *percent* saving of fuel is small, because the available heat to heat the contents and the exhaust gases is large. This relation is shown in Figure 2.4-1 for natural gas, 2.4-2 for #2 fuel oil, and 2.4-3 for #6 fuel oil. All data is based upon 10% excess air being utilized.

The percent of fuel saved is found by entering the proper chart for the given

8 INDUSTRIAL AND COMMERCIAL HEAT RECOVERY SYSTEMS

% Fuel saved by pre-heating combustion air using natural gas, 10% XSAir

t Combustion air temperature, F

t_e Furnace gas exit temperature, F	600	700	800	900	1000	1100	1200	1300	1400	1500	1600	1800	2000	2200
1000	13.4	15.5	17.6	19.6	—	—	—	—	—	—	—	—	—	—
1100	13.8	16.0	18.2	20.2	22.2	—	—	—	—	—	—	—	—	—
1200	14.3	16.6	18.7	20.9	22.9	24.8	—	—	—	—	—	—	—	—
1300	14.8	17.1	19.4	21.5	23.6	25.6	27.5	—	—	—	—	—	—	—
1400	15.3	17.8	20.1	22.3	24.4	26.4	28.4	30.2	—	—	—	—	—	—
1500	16.0	18.5	20.8	23.1	25.3	27.3	29.3	31.2	33.0	—	—	—	—	—
1600	16.6	19.2	21.6	24.0	26.2	28.3	30.3	32.2	34.1	35.8	—	—	—	—
1700	17.4	20.0	22.5	24.9	27.2	29.4	31.4	33.4	35.3	37.0	38.7	—	—	—
1800	18.2	20.9	23.5	26.0	28.3	30.6	32.7	34.6	36.5	38.3	40.1	—	—	—
1900	19.1	21.9	24.6	27.1	29.6	31.8	34.0	36.0	37.9	39.7	41.5	44.7	—	—
2000	20.1	23.0	25.8	28.4	30.9	33.2	35.4	37.5	39.4	41.3	43.0	46.3	—	—
2100	21.2	24.3	27.2	29.9	32.4	34.8	37.0	39.1	41.1	43.0	44.7	48.0	51.0	—
2200	22.5	25.7	28.7	31.5	34.1	36.5	38.8	40.9	42.9	44.8	46.6	49.9	52.8	—
2300	24.0	27.3	30.4	33.3	36.0	38.5	40.8	42.9	45.0	46.9	48.7	52.0	54.9	57.5
2400	25.7	29.2	32.4	35.3	38.1	40.6	43.0	45.2	47.2	49.2	51.0	54.2	57.1	59.7
2500	27.7	31.3	34.7	37.7	40.5	43.1	45.5	47.7	49.8	51.7	53.5	56.8	59.6	62.2
2600	30.1	33.9	37.3	40.5	43.4	46.0	48.4	50.6	52.7	54.6	56.4	59.6	62.4	64.9
2700	33.0	37.0	40.6	43.8	46.7	49.4	51.8	54.0	56.1	58.0	59.7	62.8	65.5	67.9
2800	36.7	40.8	44.5	47.8	50.8	53.4	55.8	58.0	60.0	61.9	63.5	66.5	69.1	71.3
2900	41.4	45.7	49.5	52.8	55.7	58.4	60.7	62.8	64.7	66.4	68.0	70.8	73.2	75.2
3000	47.9	52.3	56.0	59.3	62.1	64.6	66.7	68.7	70.4	72.0	73.5	75.9	78.0	79.8
3100	57.3	61.5	65.0	68.0	70.5	72.7	74.6	76.2	77.7	79.0	80.2	82.2	83.8	85.2
3200	72.2	75.6	78.3	80.4	82.2	83.7	85.0	86.1	87.1	87.9	88.7	89.9	90.9	91.8

Figure 2.4-1. Percent fuel saved by pre-heating combustion air using natural gas. (*Courtesy North American Mfg. Co., Cleveland, OH.*)

AMOUNT OF HEAT RECOVERED

% Fuel saved with #2 fuel oil, 10% XSAir / t_f, Furnace gas exit temperature, F	\ t Combustion air temperature, F													
	600	700	800	900	1000	1100	1200	1300	1400	1500	1600	1800	2000	2200
1000	13.4	15.6	17.7	19.7	21.7	—	—	—	—	—	—	—	—	—
1100	13.9	16.1	18.3	20.3	22.3	—	—	—	—	—	—	—	—	—
1200	14.3	16.6	18.8	20.9	23.0	—	—	—	—	—	—	—	—	—
1300	14.9	17.2	19.5	21.6	23.7	25.7	27.6	29.4	—	—	—	—	—	—
1400	15.4	17.8	20.1	22.3	24.5	26.5	28.4	30.3	32.0	—	—	—	—	—
1500	16.0	18.5	20.9	23.1	25.3	27.4	29.3	31.2	33.0	34.8	—	—	—	—
1600	16.6	19.2	21.7	24.0	26.2	28.3	30.3	32.3	34.1	35.8	37.5	—	—	—
1700	17.3	20.0	22.5	24.9	27.2	29.4	31.4	33.4	35.2	37.0	38.7	—	—	—
1800	18.1	20.9	23.5	25.9	28.3	30.5	32.6	34.6	36.5	38.3	40.0	43.2	—	—
1900	19.0	21.9	24.5	27.1	29.5	31.7	33.9	35.9	37.8	39.6	41.4	44.6	—	—
2000	20.0	22.9	25.7	28.3	30.7	33.1	35.2	37.3	39.3	41.1	42.9	46.1	49.0	—
2100	21.0	24.1	27.0	29.7	32.2	34.6	36.8	38.9	40.9	42.7	44.5	47.8	50.7	—
2200	22.3	25.4	28.4	31.2	33.8	36.2	38.5	40.6	42.6	44.5	46.3	49.6	52.5	55.2
2300	23.6	26.9	30.0	32.9	35.5	38.0	40.3	42.5	44.5	46.4	48.2	51.5	54.4	57.1
2400	25.2	28.7	31.9	34.8	37.5	40.1	42.4	44.6	46.7	48.6	50.4	53.7	56.6	59.2
2500	27.1	30.7	34.0	37.0	39.8	42.4	44.8	47.0	49.1	51.0	52.8	56.0	58.9	61.5
2600	29.3	33.1	36.5	39.6	42.5	45.1	47.5	49.7	51.8	53.7	55.5	58.7	61.5	64.0
2700	31.9	35.9	39.4	42.6	45.5	48.2	50.6	52.8	54.9	56.8	58.5	61.7	64.4	66.8
2800	35.2	39.3	42.9	46.2	49.1	51.8	54.2	56.4	58.4	60.3	62.0	65.1	67.7	70.0
2900	39.3	43.6	47.3	50.6	53.6	56.2	58.6	60.7	62.7	64.5	66.1	69.0	71.4	73.6
3000	44.8	49.2	52.9	56.2	59.1	61.6	63.9	65.9	67.8	69.4	70.9	73.6	75.8	77.7
3100	52.4	56.5	60.4	63.5	66.2	68.5	70.6	72.4	74.0	75.5	76.8	79.1	80.9	82.5
3200	63.8	67.8	71.0	73.6	75.9	77.8	79.4	80.8	82.1	83.2	84.1	85.8	87.2	88.3

Figure 2.4-2. Percent fuel saved by pre-heating combustion air using #2 fuel oil. (*Courtesy North American Mfg. Co., Cleveland, OH.*)

10 INDUSTRIAL AND COMMERCIAL HEAT RECOVERY SYSTEMS

t_2, Combustion air temperature, F

% Fuel saved with #6 fuel oil, 10% XSAir / t_1, Furnace gas exit temperature, F	600	700	800	900	1000	1100	1200	1300	1400	1500	1600	1800	2000	2200
1000	12.8	14.9	17.0	18.9	20.8	—	—	—	—	—	—	—	—	—
1100	13.2	15.4	17.4	19.4	21.4	—	—	—	—	—	—	—	—	—
1200	13.6	15.9	18.0	20.0	22.0	23.9	25.7	27.4	29.1	30.6	—	—	—	—
1300	14.1	16.4	18.5	20.6	22.6	24.5	26.4	28.1	29.8	31.5	—	—	—	—
1400	14.6	16.9	19.1	21.3	23.3	25.3	27.2	29.0	30.7	32.4	—	—	—	—
1500	15.1	17.5	19.8	22.0	24.1	26.1	28.0	29.8	31.6	33.3	34.9	37.9	40.8	43.4
1600	15.7	18.2	20.5	22.7	24.9	26.9	28.9	30.7	32.5	34.3	35.9	39.0	41.8	44.4
1700	16.3	18.9	21.3	23.6	25.8	27.9	29.9	31.8	33.6	35.3	37.0	40.1	43.0	45.6
1800	17.0	19.6	22.1	24.5	26.7	28.9	30.9	32.8	34.7	36.5	38.1	41.3	44.2	46.8
1900	17.8	20.5	23.0	25.5	27.8	30.0	32.0	34.0	35.9	37.7	39.4	42.6	45.5	48.1
2000	18.6	21.4	24.1	26.6	28.9	31.2	33.3	35.3	37.2	39.0	40.7	43.9	46.9	49.5
2100	19.5	22.4	25.2	27.7	30.2	32.5	34.6	36.7	38.6	40.4	42.2	45.4	48.4	51.0
2200	20.5	23.6	26.4	29.0	31.5	33.9	36.1	38.2	40.1	42.0	43.8	47.0	50.0	52.6
2300	21.7	24.9	27.8	30.5	33.1	35.5	37.7	39.8	41.8	43.7	45.5	48.8	51.7	54.4
2400	23.0	26.3	29.3	32.1	34.8	37.2	39.5	41.7	43.7	45.6	47.4	50.7	53.6	56.3
2500	24.6	27.9	31.1	34.0	36.7	39.2	41.5	43.7	45.8	47.7	49.5	52.8	55.7	58.3
2600	26.3	29.9	33.1	36.1	38.9	41.5	43.8	46.0	48.1	50.0	51.8	55.1	58.0	60.6
2700	28.4	32.1	35.5	38.6	41.4	44.0	46.4	48.6	50.7	52.6	54.4	57.7	60.5	63.0
2800	30.9	34.8	38.3	41.5	44.4	47.0	49.4	51.7	53.7	55.6	57.4	60.6	63.4	65.8
2900	34.0	38.1	41.7	44.9	47.9	50.6	53.0	55.2	57.2	59.1	60.8	63.9	66.6	68.9
3000	38.0	42.2	45.9	49.2	52.2	54.8	57.2	59.4	61.4	63.2	64.8	67.8	70.3	72.5
3100	43.2	47.6	51.4	54.7	57.6	60.2	62.5	64.5	66.4	68.1	69.6	72.3	74.6	76.6
3200	50.6	55.0	58.7	61.8	64.6	67.0	69.1	71.0	72.6	74.1	75.5	77.8	79.8	81.5

Figure 2.4-3. Percent fuel saved by pre-heating combustion air using #6 fuel oil. (*Courtesy North American Mfg. Co., Cleveland, OH.*)

combustion air pre-heat and exhaust gas temperatures. (Methods for determining the pre-heat temperature are given in Chapter 5.) The amount of energy saving, H, is the amount currently used, u, times the percentage saving, F, divided by 100:

$$H = \frac{Fu}{(100)}. \qquad (2\text{-}7)$$

2.5 CALCULATION OF COST SAVINGS

The cost saving from a heat recovery system is the cost of purchased energy avoided by using the recovered heat. The saving is based only on the amount of recovered heat used, not on the amount of heat available from the source. The cost saving is credited for any efficiency of the heat source being replaced, since the recovered heat is already available to the use as heat.

The cost saving, C, resulting from the recovery and use of heat, H, for t operating hours is

$$C = \frac{100HECt}{E}. \qquad (2\text{-}8)$$

The energy cost, EC, and the efficiency, E, refer to the heat-producing equipment being supplemented or replaced by the heat recovery equipment. The value of E increases C because the heat, H, does not have to be produced in the heat-producing equipment, and is credited with its efficiency, E.

When the time unit for C and t is one year, C is the annual cost saving and t is the number of hours per year of heat recovery system operation. The energy cost, EC, is determined by the current energy prices and the type of energy being utilized. To convert from market units to energy units, the following conversion factors are used.

$$\text{kilowatt-hours} \times 3413 = \text{Btu}$$

$$\text{cubic feet natural gas} \times 1020 = \text{Btu}$$

$$\text{therms natural gas} \times 100{,}000 = \text{Btu}$$

$$\text{gallons \#2 oil} \times 135{,}000 = \text{Btu}$$

$$\text{gallons \#4 oil} \times 145{,}000 = \text{Btu}$$

$$\text{gallons \#6 oil} \times 155{,}000 = \text{Btu}$$

$$\text{gallons propane} \times 91{,}500 = \text{Btu}$$

12 INDUSTRIAL AND COMMERCIAL HEAT RECOVERY SYSTEMS

These conversion factors may vary according to fuel source, but are sufficiently accurate for calculation purposes. The energy cost, EC, is the billed cost in dollars, B, divided by the amount purchased, AP, times its appropriate conversion factor, N:

$$EC = \frac{B}{APN}. \tag{2-9}$$

Equations 2-8 and 2-9 show that heat recovery system evaluation should not be limited to applications recovering large amounts of recovered heat. Expensive energy sources, such as electric heat, can provide a large cost saving for a small amount of recovered heat. Similarly, many hours of operation can produce a large annual cost saving from a small amount of recovered heat.

3
HEAT RECOVERY SURVEY

A heat recovery survey provides the data required for the design of the heat recovery system. The survey locates all sources of recoverable heat and all uses for recovered heat within the property limits. Both sources and uses are measured for flow, temperature, pressure, composition, contaminants, and operating cycles. The sources and uses are combined into possible heat recovery systems, whose installation cost and financial return can be calculated. The final choice of heat recovery system is a choice of the best financial return balanced against the least risk and maintenance costs. These factors are discussed in succeeding chapters.

3.1 LOCATION OF SOURCES AND USES OF RECOVERED HEAT

A heat recovery survey begins with a walking tour of the facility. Each source and use of heat is noted, including gas, liquid, and vapor. The following list of information should accompany each source or use.

Composition   Condition of Ducts, Pipes, Equipment
Location   Contaminants
Size of Supply and Exhaust   Installation Problems
Operating Cycle   Supply and Exhaust Operating Conditions

Some of the typical sources are listed below.

14 INDUSTRIAL AND COMMERCIAL HEAT RECOVERY SYSTEMS

Air

Roof Ventilation Exhaust	Laboratory Exhaust
Oven/Dryer Exhaust	Dust Collector Exhaust
Transformer Vault Exhaust	Kitchen Hood Exhaust
Boiler Room Exhaust	Cooling Air Exhaust

Combustion Products

Boilers	Air Heaters	Bakery Ovens
Furnaces	Space Heaters	Textile Dryers
Ovens	Paper Dryers	Incinerators
Engines	Gas Turbines	Baking Ovens

Liquids

Textile Dye Works	Boiler Blow-down	Compressor Coolant
Paper Mills	Boiler Condensate	Heat Transfer Liquids
Food Preparation	Process Cooling Water	Food Preparation

Vapors

Steam Discharge	Cooking Kettles	Distillation Processes
Evaporation Processes	Solvent Vapor Exhaust	Paper Making

Some of the typical uses are:

Fuel Burning

Ovens	Space Heaters	Tank Heaters
Furnaces	Dryers	Process Equipment
Boilers	Air Makeup Units	Heat Transfer Fluid Heaters

Heated Air

Dryers	Food Processors
Air Makeup Units	Calciners
Ovens	Building Heat

Vapors and Steam

Steam Dryers	Absorption Air Conditioners
Boiler Feed Water	Solvent Cleaners
Process Kettles	Hot Water Heaters

HEAT RECOVERY SURVEY 15

3.2 COMPOSITION

The composition of the supply and exhaust is required for heat recovery equipment selection and for heat recovery calculation. The composition indicates the presence of corrosive conditions and the possibility of condensation in the heat recovery equipment. Composition also provides data for determining the proper values of the heat of vaporization or condensation, h_{cos}, the specific heat C_p, and the density, d, at the operating conditions. For gases other than air, and liquids other than water, this data is available from chemical handbooks and suppliers' literature. The operating conditions are determined by methods described later in this chapter.

3.3 LOCATION

Each source and use is located on a scale plan of the facility, along with obstructions and structural details. Separation distance between pairs of sources and uses will allow choice of the pair with the least piping or ducting, and estimation of piping or ducting cost. Location of obstructions is necessary if special ducting or piping is required to avoid the obstruction. Structural information will help in planning building modifications to accommodate the size and weight of the heat recovery equipment.

3.4 SIZE OF SUPPLY AND EXHAUST SYSTEMS

Diameter, height, orientation, shape, and any other special description of the supply and exhaust systems is noted on the plan to establish ducting and piping connection requirements, and installation requirements for the heat recovery equipment. These details will also aid in planning measurements of the source and use.

3.5 OPERATING CYCLE

The source and use operating cycles establish the operating time used in calculating the cost saving (Equation 2-8). The number of hours/day, days/week, and weeks/year provide the annual operating hours of use. Operating times for sources and uses should be recorded to allow choice of maximum common operating time when pairing a source and use.

Operating cycle also refers to variations in operating conditions during the hours of operation. Combustion equipment may alternate between high-fire and low-fire operation. Batch ovens and dryers will begin a cycle at low-temperature, high-humidity conditions, and finish the cycle at high-temperature, low-humidity conditions. Building cooling use will reach a peak during full occupancy, and diminish during periods of low occupancy. Operating cycle data are required to establish the peak, average and minimum source and use operating conditions. Average operating conditions are determined from operating cycle data to calculate the heat recovered and cost saving. Peak and minimum operating conditions are used for selecting heat recovery equipment, components, controls, and materials of construction where extreme conditions may affect the heat recovery system design and performance.

Heat recovery survey notes should list the operating times, days of the week, weeks of the year, operating cycles of the source, and use and variation of the source and use conditions during the operating cycle.

3.6 CONDITION OF DUCTS, PIPES, EQUIPMENT

The condition of the ducts, pipes, and equipment is recorded in the heat recovery survey. It provides an indication of any corrosive or contaminated operating condition. Areas of rust, holes in ducting, stains or discoloration, and abraded areas are all danger signs for corrosion, condensation, or abrasive particulates. Any accumulation on the inside of ducts, stacks, or pipes must be carefully examined, because the same accumulation will occur in the heat recovery equipment and degrade its performance.

The condition of supporting structure, ducts, pipes, and equipment is important for estimating heat recovery system costs. Items requiring replacement, reinforcement, or repair should be included in the heat recovery survey notes.

3.7 CONTAMINANTS

Contaminants in the source can lead to failure of the heat recovery system, unless the system is properly designed to handle the contamination. Several types of contaminants may be present in the exhaust and supply, which are troublesome. Their presence should be recorded in the heat recovery survey notes.

Solid particulates can abrade or plug heat recovery equipment. Hard, abrasive materials will wear out elbows, fans, pumps, or wherever a flow direction is changed. Any particulate which builds up will clog flow passages in heat recovery equipment. This is true of strings, fluffy or fibrous textiles, sticky organics, or wet powders. The last items are particularly difficult to handle if allowed to dry, because they form a hard deposit which must be chipped off.

HEAT RECOVERY SURVEY   17

During the survey, tests can be made with various detergents to find an effective cleaning agent for any existing deposits. Most detergent manufacturers will supply samples for tests. The detergent can be used in washing equipment supplied with the heat recovery equipment, or for cleaning filters included in the heat recovery system. Self-curing compounds, like epoxies and certain inks, cannot be removed by washing. Throw-away filters should be used to avoid plugging the heat recovery equipment. In extreme cases where filter replacement is too costly, the application of heat recovery should be abandoned.

Moisture is a contaminant which can be handled by proper design. It may occur as part of the exhaust or supply, or it may be formed as a condensate when they are cooled. Organic vapors may also condense when cooled. If condensation can occur in the heat recovery system, exhaust and supply dewpoints should be measured. Frequently, the moisture or condensate will be acidic because of the presence of acid-forming gases in the exhaust stream. When dissolved in the moisture or condensate, sulfuric, nitric, and hydrochloric acid form. Special materials of construction will be required to resist the acid attack. A sample of the condensate can be collected from an ice bath in the exhaust stream, and tested for acidity (pH) and composition.

Greases are a common contaminant in kitchen exhausts. They collect on fans, ducts, and heat recovery equipment, but can be removed by spraywashing the equipment.

3.8 INSTALLATION PROBLEMS

The heat recovery survey should note any geographic or construction features of the facility that will affect the cost or the design of the heat recovery system. Geographic features include ponds, streets, obstructing buildings, electric poles, etc. Construction features include reinforced concrete walls, low ceilings, electric conduits and pipes in a congested area, roof supports, size of access doors, etc. The information contained in this section of the survey notes will be useful in selecting equipment size and type to fit into an available location, and to determine any rigging problems.

3.9 SUPPLY AND EXHAUST OPERATING CONDITIONS

The supply fluid coming into the use and the exhaust fluid from the source should be measured during the heat recovery survey. Methods of measurement are discussed later in this chapter. Care is taken during the measurements to establish the source and use methods of operation: peak, minimum, average, or even if the source or use are shut-down. Operating personnel

18 INDUSTRIAL AND COMMERCIAL HEAT RECOVERY SYSTEMS

should be notified not to change the source or use operation during the measurements.

3.10 HEAT RECOVERY MEASUREMENTS

Calculation of the amount of recoverable heat, amount of useful recovered heat, and sizing of the heat recovery equipment requires knowledge of the flow, temperature, and pressure of the supply and exhaust. These data may be available from equipment or process specifications, or they may have to be measured. If the process or equipment has been changed since its installation, the specifications may be invalid, requiring measurements. This section reviews the methods of measuring flow, temperature, and pressure. Methods for measuring composition, contaminants, and dewpoint are not discussed, and can be found in chemical handbooks.

3.10.1 Pressure

Pressure is used to correct for standard conditions for air or gas in Equation 2-5, to measure pressure drop across heat recovery equipment, and to verify the presence of adequate pressure for equipment operation.

Only a single measurement is necessary because pressure is constant across a flowing stream or in a storage tank. At moderate temperatures and pressures, pressure is measured by drilling a 1/16-in.-diameter hole in a smooth-flowing portion of ducting or piping. A plastic tube connected to one side of a U-tube manometer is held tightly against the hole, and the height difference, Δh, between the two legs of the manometer is read. The static pressure, P, is the difference in height, Δh, times the density, d, of the liquid in the U-tube:

$$P = d \, \Delta h. \tag{3-1}$$

If d is in $lb/in.^3$ and Δh is in in., P is in units of psig.

For higher temperatures, a copper tube is brazed or fitted to the duct in a screwed fitting. The plastic hose is slipped over the end of the copper tube. The tube is long enough to dissipate the duct temperature. The U-tube manometer can be filled with water, mercury, or special oils. For a manometer with a 36-in. range, the maximum pressure for measurement with mercury with a specific gravity of 13.59 is

$$P = \frac{(36)(62.4)(13.59)}{(1728)} = 17.7 \text{ psig.} \tag{3-1}$$

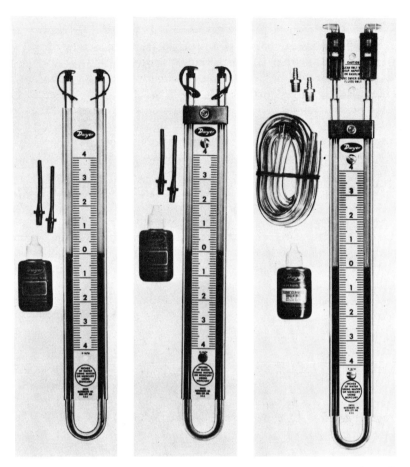

Figure 3.10.1-1. Typical U-tube manometer. (*Courtesy Dwyer Instruments, Inc., Michigan City, IN.*)

For higher pressures, a pressure gauge is screwed into the piping. A typical U-tube manometer is shown in Figure 3.10.1-1.

3.10.2. Temperature

Temperature is used to correct for standard conditions for air or gas in Equation 2-5, for calculation of heat recovery equipment performance, and to choose the proper equipment construction for operation at design temperature.

20 INDUSTRIAL AND COMMERCIAL HEAT RECOVERY SYSTEMS

Stream temperatures are measured through a small hole in the duct wall. Glass or dial thermometers are used below 750° F / 1000° F. Dial thermometers are generally used because they do not easily break and are easier to read while making the measurement. At temperatures above 500° F, the radiating heat from the stack or duct can become very uncomfortable in close quarters, and the thermometer difficult to hold. Where a cyclical process is present, such as a burner alternating between high-fire and low-fire, a slow-responding thermometer may never indicate a stable temperature. A faster-responding instrument should be used. A dial thermometer is shown in Figure 3.10.2-1. For higher temperatures, a pyrometer or a resistance thermometer should be used. They are available in ranges to over 3000° F, and have read-outs in both dial and digital form. A simple hand-held pyrometer is shown in Figure 3.10.2-2.

The tip of the pyrometer contains a thermocouple. When heated by the gas stream, the thermocouple generates a voltage, which is read on the pyrometer meter. A more accurate form contains a self-balancing potentiometer whose output is recorded on a chart. A resistance thermometer utilizes an element of known resistance/temperature characteristic in a small probe. The instrument measures the change of element resistance with a wheatstone-bridge or microelectronic circuit and converts the resistance to a temperature. These

Figure 3.10.2-1. Dial theromometer. (*Courtesy Palmer Instruments, Inc., Cincinnati, OH.*)

HEAT RECOVERY SURVEY 21

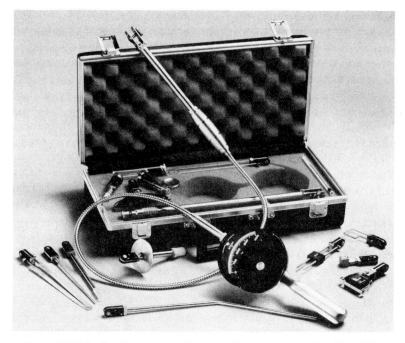

Figure 3.10.2-2. Simple pyrometer. (*Courtesy Alnor Instrument Co., Niles, IL.*)

instruments are suitable for cyclical processes because of their fast response and their recording capability. However, they are more expensive and more difficult to use than the theromometers mentioned previously.

Surface temperature may be important if the flowing temperature cannot be measured. If the surface is insulated, there are no sharp temperature changes, and good thermal contact is maintained, the surface temperature will be within a few degrees of the fluid temperature. Surface temperature can be most accurately measured with a special contact button type of thermometer or pyrometer.

Pipeline temperatures are normally measured using a well. The well holds the thermometer or temperature probe in good thermal contact with its walls, which are at liquid temperature. The well is part of the pipeline fittings.

3.10.3 Flow

Liquid flow is measured either by hand or with a flow meter. Where the liquid can be diverted into a receiver of known volume and the filling time measured, its flow can be easily calculated. Otherwise, a variety of liquid meters are used.

22 INDUSTRIAL AND COMMERCIAL HEAT RECOVERY SYSTEMS

A discussion of the various types of liquid meters is beyond the scope of this book. Reference is made to any of the chemical equipment catalogs, which contain a listing of types and their manufacturers.

Air or gas flow is most accurately measured inside a duct or stack of cross-sectional area, A, at a velocity, v, using Equation 3-2:

$$V = vA. \qquad (3\text{-}2)$$

If v is in fpm and A is in ft^2, V is in cfm.

The flow cross-section area, A, is the stack or duct area perpendicular to the flow direction. The measurement of the velocity, v, is carried out by making a "traverse" of the flow area to establish the average velocity. Detailed instructions for making a traverse and calculating the average velocity are provided in Reference 1 at the end of this chapter. The information provided in this section will, however, be adequate for most situations.

Weight flow, W, is the product of the density, d, times the volume flow, V:

$$W = dV. \qquad (3\text{-}3)$$

If d is in lb/ft^3 and V is in cfm, W is in lb/min.

A "traverse" means to measure the gas or air velocity at selected points across the flow cross-section area, A. The average of the measured velocities is used as v in Equation 3-2. The traverse points are selected to represent equal portions of the flow cross-section area. This selection equally weights the measured velocities in the averaging process. Figure 3.10.3-1 shows the distance from the wall of a circular duct for an equal area traverse. Two traverses are made, each 90° apart, to obtain the velocity distribution in each quarter of the duct. Rectangular ducts are divided into equal area portions, and each portion measured at its center. The number of traverse points is chosen to measure high- and low-velocity areas across the duct. A trial run is usually made before the traverse to establish the existing velocity profile. The traverses are commonly made with either of two types of instruments. The first, a Pitot tube, is shown in Figure 3.10.3-2.

Traverse Method	Probe immersion in Duct Diameters									
	d_1	d_2	d_3	d_4	d_5	d_6	d_7	d_8	d_9	d_{10}
6 point	0.043	0.147	0.296	0.704	0.853	0.957	–	–	–	–
8 point	0.032	0.105	0.194	0.323	0.677	0.806	0.895	0.968	–	–
10 point	0.025	0.082	0.146	0.226	0.342	0.658	0.774	0.854	0.918	0.975

Figure 3.10.3-1. Equal area traverse points. (*Courtesy Alnor Instrument Co., Niles, IL.*)

HEAT RECOVERY SURVEY 23

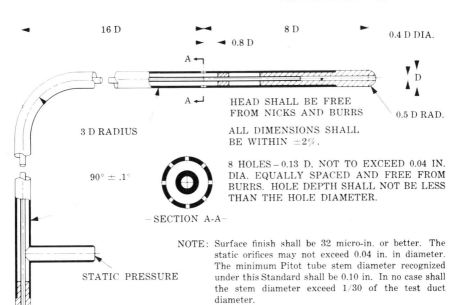

Figure 3.10.3-2. Pitot tube. (*Courtesy Air Movement and Control Association, Inc., Arlington Heights, IL.*)

The Pitot tube consists of an inner and outer tubing. The inner tubing is open only at the front of the Pitot tube. The front opening is pointed directly into the flow. The outer tubing has side holes that are perpendicular to the flow direction; they receive no input from the gas or air velocity. The inner and outer tubing are connected by flexible tubing to either side of the U-tube manometer. At low velocities, the manometer is inclined to produce accurate readings. The difference in height between the legs of the U-tube is related to the flow velocity. The velocity, v, is found from Equation 3-4, which is the definition of the relation between velocity, v, and pressure caused by velocity, ΔP, and the gravitational content, g:

$$\Delta P = \frac{dv^2}{2g}. \tag{3-4}$$

The gas density, d, is at the same condition as the flowing gas. If the pressure, ΔP, is measured as a height, Δh, in in. of water column in a U-tube manometer, ΔP can be expressed, using Equation 3-1, as

$$\frac{d_{\text{water}} \Delta h}{12} = \frac{dv^2}{2g}. \tag{3-5}$$

Equation 3-5, with appropriate values for water, becomes:

$$v = 1096 \sqrt{\frac{\Delta h}{d}} \text{ fpm}. \tag{3-6}$$

The density of air at a temperature, T, and pressure, P, is calculated from Equation 3-7:

$$d = (0.075)\left(\frac{530}{T + 460}\right)\left(\frac{P_{\text{std}} + P}{P_{\text{std}}}\right) \text{ lb/ft}^3. \tag{3-7}$$

As in Chapter 2, P_{std} is chosen in the same unit as P: 29.92 in. mercury, 14.7 psig, etc. The correction is not made when P is in the range of in. of water.

For air at standard conditions:

$$v = 4005\sqrt{\Delta h} \text{ fpm}. \tag{3-8}$$

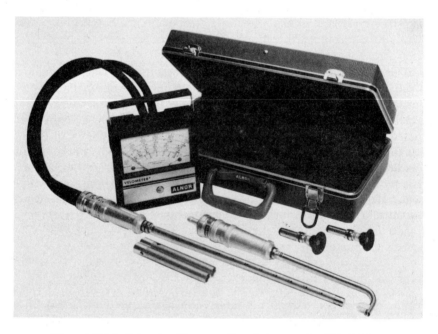

Figure 3.10.3-3. Velometer. (*Courtesy Alnor Instrument Co., Niles, IL.*)

HEAT RECOVERY SURVEY 25

Equations 3-6 and 3-8 indicate that v is proportional to the square root of Δh. When the traverse points are added together to find the average velocity, the *square root of* Δh for each point is added, the *average square root of* Δh found, and converted to v using Equation 3-6 or 3-8.

A second common type of velocity instrument is the Velometer. The Velometer is a swinging-vane instrument with a special probe and a dial face calibrated directly in fpm. Various probe bases and adjustments are available to change the instrument range from 300 to 10,000 fpm. Since the traverse is in fpm, the equal area traverse velocities are averaged directly. The meter is calibrated for standard conditions. For other temperatures, the correction of Equation 3-9 is used.

$$v = v_{read} \sqrt{\frac{T + 460}{T_{ambient} + 460}} \text{ fpm.} \qquad (3-9)$$

The flow temperature, T, and ambient temperature, $T_{ambient}$, are in degrees Fahrenheit.

3.10.4 Estimating Flow Rates

Sometimes flow rates cannot be measured because of inaccessibility, flow obstructions, lack of time, or unsafe surroundings. Several approximate methods are available for estimating flow rates. These methods are also useful for checking flow measurement.

Figure 3.10.4-1 is a typical fan performance chart. In the chart, the fan output in cfm, the fan speed in rpm, the fan brake horsepower (BHP), and the pressure rise in in. water (SP) are given. The fan speed is measured by a tachometer and the brake horsepower calculated from measured motor current and voltage. The pressure rise across the fan is measured as described in the previous section. The combination of the three provides a volume flow location on Figure 3.10.4-1. Although the fan speed can be calculated from the drive motor rpm and the ratio of drive motor pulley diameter to fan pulley diameter, belt slippage can have a considerable effect on the accuracy of the results. Wherever possible, a tachometer should be used directly on the fan shaft to measure fan speed. Liquid flow can be estimated in the same manner. A typical pump performance diagram is shown in Figure 3.10.4-2.

The diagram corresponds to a particular pump speed. The horizontal curved lines refer to an impeller diameter, and the rounded lines are horsepower requirements. The total head in feet of water is measured from the pump lift or with a pressure gauge. Knowing the pump speed, head, and impeller diameter, the flow rate is read at the horizontal scale. If the pump head cannot be measured, the pump horsepower can be calculated from

CFM	OV	1/4" SP		3/8" SP		1/2" SP		3/4" SP		1" SP		1 1/2" SP		2" SP		2 1/2" SP		3" SP		3 1/2" SP	
		RPM	BHP	RPM	BHP	RPM	BHP	RPM	BHP	RPM	BHP	RPM	BHP	RPM	BHP	RPM	BHP	RPM	BHP	RPM	BHP
4135	800	334	0.25																		
4652	900	357	0.31																		
5169	1000	382	0.37																		
5686	1100	408	0.45	373	0.34	411	0.44														
6203	1200	434	0.53	392	0.41	427	0.51														
6720	1300	462	0.63	414	0.48	445	0.59														
7237	1400	490	0.75	437	0.56	466	0.69	482	0.66												
7754	1500	517	0.87	462	0.66	488	0.79	492	0.74	555	1.01										
8271	1600	546	1.02	487	0.77	512	0.91	506	0.84	564	1.11										
8788	1700	575	1.17	513	0.89	537	1.04	522	0.94	577	1.22	679	1.83								
9305	1800	604	1.35	541	1.03	563	1.19	541	1.06	591	1.35	688	1.98	777	2.68						
9822	1900	632	1.54	568	1.18	588	1.34	561	1.21	608	1.51	700	2.16	786	2.88						
10339	2000	662	1.76	595	1.34	616	1.53	582	1.35	627	1.66	714	2.35	796	3.08	874	3.89				
10856	2100	692	2.01	623	1.53	642	1.72	605	1.51	648	1.85	729	2.55	808	3.32	883	4.14	956	5.04		
11373	2200	721	2.25	651	1.73	669	1.93	629	1.69	669	2.04	746	2.78	822	3.57	894	4.42	963	5.31		
11890	2300	751	2.53	680	1.96	697	2.16	654	1.88	692	2.26	765	3.02	837	3.84	906	4.71	974	5.63	1030	6.27
12407	2400	781	2.83	709	2.21	726	2.42	680	2.11	715	2.48	786	3.29	853	4.13	920	5.03	986	5.97		
12924	2500	810	3.16	738	2.47	755	2.71	705	2.32	740	2.74	806	3.57	872	4.45	936	5.37	998	6.34		
13441	2600	841	3.52	767	2.76	783	2.99	732	2.59	764	3.01	829	3.88	891	4.78	952	5.73	1012	6.73	1038	6.61
	2700			796	3.07	812	3.32	759	2.86	790	3.29	851	4.21	911	5.14	970	6.12	1027	7.13	1048	6.97
				826	3.41	841	3.66	786	3.16	816	3.61	876	4.57	933	5.54	989	6.53	1045	7.58	1058	7.34
	2800	901	4.31																		
14475																					
				856	3.78	869	4.03	813	3.47	842	3.94	898	4.91	955	5.95	1009	6.97	1063	8.05	1071	7.76
										869	4.31	924	5.32	977	6.36	1030	7.45	1081	8.54	1084	8.19
										896	4.68	949	5.74	1000	6.81	1051	7.91	1100	9.04	1100	8.66
				915	4.58	929	4.87	897	4.56	924	5.11	975	6.21	1024	7.31	1073	8.43	1122	9.61	1115	9.14
								954	5.43	980	6.01	1027	7.16	1074	8.35	1120	9.56	1165	10.8	1132	9.67
																				1151	10.2
																				1170	10.8
																				1210	12.0

CFM	OV	4" SP		4½" SP		5" SP		5½" SP		6" SP		6½" SP		7" SP		7½" SP		8" SP		8½" SP	
		RPM	BHP	RPM	BHP	RPM	BHP	RPM	BHP	RPM	BHP	RPM	BHP	RPM	BHP	RPM	BHP	RPM	BHP	RPM	BHP
9305	1800	1109	8.01	1167	9.07																
9822	1900	1118	8.41	1176	9.51	1232	10.6														
10339	2000	1128	8.83	1185	9.97	1238	11.1														
10856	2100	1140	9.29	1194	10.4	1248	11.6														
11373	2200	1154	9.79	1205	10.9	1257	12.1	1292	12.3												
11890	2300	1167	10.2	1218	11.4	1269	12.7														
12407	2400	1183	10.8	1232	12.0	1282	13.3	1300	12.8	1349	14.1										
12924	2500	1200	11.4	1248	12.6	1296	13.9	1308	13.3	1357	14.6										
13441	2600	1217	12.0	1263	13.3	1310	14.5	1318	13.9	1367	15.2										
14475	2800	1254	13.3	1298	14.6	1341	16.0	1330	14.6	1377	15.9	1406	15.9	1461	18.0	1505	19.3				
15509	3000	1296	14.8	1336	16.1	1376	17.5	1342	15.2	1387	16.5	1413	16.5								
16543	3200	1339	16.3	1377	17.8	1416	19.2	1355	15.9	1399	17.2										
17577	3400	1383	18.0	1420	19.5	1457	21.1	1385	17.3	1427	18.7	1423	17.2	1469	18.6	1513	20.0	1557	21.5	1606	23.7
18611	3600	1431	19.9	1467	21.5	1501	23.0	1419	19.0	1458	20.4	1433	17.9	1477	19.3	1521	20.7	1563	22.2	1615	24.6
19645	3800	1479	21.9	1513	23.5	1547	25.2	1454	20.7	1492	22.2	1444	18.6	1487	20.0	1531	21.5	1572	23.0		
20679	4000	1529	24.0	1562	25.8	1593	27.5	1494	22.6	1530	24.2	1470	20.2	1510	21.6	1553	23.2	1593	24.7	1632	26.2
								1536	24.6	1570	26.3	1498	21.9	1537	23.4	1577	24.9	1616	26.5	1654	28.1
								1579	26.8	1612	28.5	1531	23.8	1570	25.4	1607	27.0	1644	28.6	1680	30.2
								1626	29.3	1658	31.1	1568	25.8	1601	27.3	1640	29.1	1674	30.7	1709	32.4
												1606	28.0	1638	29.6	1675	31.4	1707	33.0	1740	34.8
												1647	30.3	1677	32.0	1712	33.8	1742	35.5	1777	37.4
												1689	32.8	1721	34.7	1751	36.4	1783	38.3	1813	40.1

Figure 3.10.4-1. Typical fan performance chart. (*Courtesy The New York Blower Co., Willowbrook, IL.*)

28 INDUSTRIAL AND COMMERCIAL HEAT RECOVERY SYSTEMS

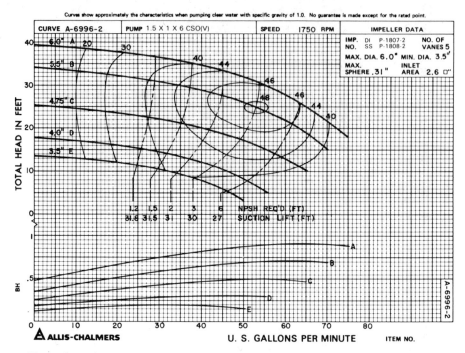

Figure 3.10.4-2. Typical pump performance diagram. (*Courtesy Allis-Chalmers Corp., Milwaukee, WI.*)

measured current and voltage. The flow rate is read directly below the pump impeller size and horsepower intersection.

Combustion exhaust flow rates can be estimated from the burner input. For theoretically correct combustion,

1 cubic foot of natural gas produces 11 ft^3 of combustion product
1 ft^3 of propane produces 25 ft^3 of combustion product
1 gallon of fuel oil produces 1450 ft^3 of combustion product

Most commercial and industrial burners are adjusted to operate with excess air. The presence of the excess air insures complete combustion of the fuel, especially where pockets of fuel-rich mixture may develop. The normal range of excess air will run from 10% to 50%, although it frequently is much higher. An estimate of combustion exhaust gas volume should include an excess air measurement. Excess air volume is measured by knowing the oxygen or carbon dioxide volume present, and then converting it to air volume using Figure 3.10.4-3. The excess air volume is added to the theoretical combustion exhaust flow to arrive at total exhaust flow. Oxygen and carbon dioxide

HEAT RECOVERY SURVEY

						% Excess air						
		0	10	20	40	60	80	100	200	400	1000	
						% Theoretical air						
%O₂ (standard type)⊙		100	110	120	140	160	180	200	300	500	1100	
%CO₂ (bold)						φ, Equivalence ratio						
a/f ratio* (italic)		1.00	0.91	0.83	0.71	0.62	0.56	0.50	0.33	0.20	0.09	
Fuels	Natural gas (Birmingham)	0	2.09	3.80	6.43	8.36	9.83	11.00	14.42	17.07	19.19	
		11.74‡	**10.57**	**9.61**	**8.14**	**7.05**	**6.22**	**5.57**	**3.65**	**2.16**	**0.97**	
		9.44	*10.38*	*11.33*	*13.22*	*15.10*	*16.99*	*18.88*	*28.32*	*47.2*	*103.8*	
	Blast furnace gas	0	0.89	1.71	3.16	4.41	5.50	6.45	9.86	13.40	17.09	
		25.51‡	**24.42**	**23.43**	**21.65**	**20.13**	**18.81**	**17.65**	**13.49**	**9.17**	**4.68**	
		0.68	*0.75*	*0.82*	*0.95*	*1.09*	*1.23*	*1.36*	*2.04*	*3.4*	*7.5*	
	Producer gas (W-G, bituminous)	0	1.35	2.54	4.54	6.14	7.46	8.56	12.15	15.37	18.28	
		18.48‡	**17.28**	**16.23**	**14.47**	**13.05**	**11.89**	**10.92**	**7.75**	**4.90**	**2.33**	
		1.30	*1.43*	*1.56*	*1.82*	*2.08*	*2.34*	*2.60*	*3.90*	*6.5*	*14.3*	
	Coke oven gas (by-product)	0	2.10	3.82	6.45	8.39	9.86	11.03	14.44	17.09	19.20	
		10.82‡	**9.73**	**8.84**	**7.48**	**6.48**	**5.72**	**5.11**	**3.35**	**1.98**	**0.89**	
		5.42	*5.97*	*6.51*	*7.59*	*8.68*	*9.76*	*10.85*	*16.27*	*27.1*	*59.7*	
	Propane (natural)	0	2.06	3.75	6.36	8.28	9.75	10.92	14.35	17.02	19.16	
		13.69‡	**12.34**	**11.24**	**9.53**	**8.27**	**7.31**	**6.55**	**4.30**	**2.55**	**1.15**	
		23.78	*26.16*	*28.53*	*33.29*	*38.04*	*42.80*	*47.56*	*71.33*	*118.9*	*261.6*	
	Butane (refinery)	0	2.05	3.74	6.34	8.26	9.73	10.90	14.33	17.01	19.16	
		13.99‡	**12.62**	**11.49**	**9.75**	**8.46**	**7.48**	**6.70**	**4.41**	**2.61**	**1.18**	
		30.65	*33.71*	*36.78*	*42.91*	*49.04*	*55.17*	*61.30*	*91.95*	*153.2*	*337.1*	

* ft³ air/ft³ fuel, m³ air/m³ fuel, or any ratio of volumes in consistant units.
‡ Ultimate %CO₂.
⊙ Dry basis. See Handbook Supplement 168-SF1, 2, 3 for conversions to wet basis.

Figure 3.10.4-3a. Percent of excess air from percent measured oxygen or carbon dioxide for various fuels. (*Courtesy North American Mfg. Co., Combustion Handbook, Cleveland, OH.*)

30   INDUSTRIAL AND COMMERCIAL HEAT RECOVERY SYSTEMS

Fuels		% O₂ / %CO₂ / a/f ratio†	0 / 100 / 1.00	10 / 110 / 0.91	20 / 120 / 0.83	40 / 140 / 0.71	% Excess air: 60 / % Theoretical air: 160 / Φ, Equivalence ratio: 0.62	80 / 180 / 0.56	100 / 200 / 0.50	200 / 300 / 0.33	400 / 500 / 0.20	1000 / 1100 / 0.09
#1	Distillate oil		0 / 15.40‡ / 191.57	2.04 / 13.90 / 210.73	3.71 / 12.66 / 229.89	6.31 / 10.76 / 459.78	8.22 / 9.35 / 306.52	9.69 / 8.27 / 344.83	10.86 / 7.41 / 383.15	14.29 / 4.88 / 574.72	16.98 / 2.90 / 957.9	19.15 / 1.31 / 2107.3
#2	Distillate oil		0 / 15.68‡ / 189.04	2.03 / 14.15 / 207.95	3.70 / 12.90 / 226.85	6.29 / 10.96 / 453.70	8.20 / 9.53 / 302.47	9.67 / 8.43 / 340.27	10.84 / 7.56 / 378.08	14.28 / 4.98 / 567.12	16.97 / 2.96 / 945.2	19.14 / 1.33 / 2079.5
#4	Blended oil		0 / 15.83‡ / 184.60	2.02 / 14.30 / 203.06	3.69 / 13.04 / 221.52	6.28 / 11.08 / 443.05	8.19 / 9.64 / 295.37	9.66 / 8.53 / 332.29	10.82 / 7.64 / 369.21	14.26 / 5.04 / 553.81	16.96 / 3.00 / 923.0	19.13 / 1.35 / 2030.6
#5	Residual oil		0 / 16.31‡ / 183.19	2.02 / 14.74 / 201.51	3.68 / 13.44 / 219.83	6.26 / 11.43 / 439.66	8.16 / 9.94 / 293.11	9.63 / 8.80 / 329.75	10.80 / 7.89 / 366.39	14.24 / 5.21 / 549.58	16.95 / 3.10 / 916.0	19.13 / 1.40 / 2015.1
#6	Residual oil		0 / 16.79‡ / 175.99	2.00 / 15.18 / 193.59	3.66 / 13.86 / 211.19	6.23 / 11.79 / 422.38	8.13 / 10.26 / 281.59	9.60 / 9.09 / 316.79	10.76 / 8.15 / 351.99	14.21 / 5.38 / 527.98	16.93 / 3.21 / 880.0	19.12 / 1.45 / 1935.9
	Bituminous coal		0 / 18.53‡ / 141.59	1.97 / 16.79 / 155.74	3.59 / 15.35 / 169.90	6.13 / 13.10 / 339.81	8.02 / 11.42 / 226.54	9.49 / 10.13 / 254.85	10.65 / 9.10 / 283.17	14.12 / 6.03 / 424.76	16.86 / 3.60 / 707.9	19.08 / 1.63 / 1557.4
	Anthracite coal		0 / 19.92‡ / 130.20	1.93 / 18.08 / 143.22	3.54 / 16.55 / 156.24	6.06 / 14.15 / 312.48	7.94 / 12.36 / 208.32	9.39 / 10.97 / 234.36	10.56 / 9.87 / 260.40	14.03 / 6.56 / 390.59	16.80 / 3.92 / 651.0	19.05 / 1.78 / 1432.2
	Coke		0 / 20.50‡ / 132.21	1.92 / 18.62 / 145.43	3.52 / 17.06 / 158.65	6.02 / 14.60 / 317.30	7.89 / 12.77 / 211.53	9.35 / 11.34 / 237.98	10.51 / 10.20 / 264.42	13.99 / 6.79 / 396.63	16.77 / 4.07 / 661.0	19.04 / 1.85 / 1454.3

† ft³ air/lb fuel.
‡ Ultimate %CO₂.
○ Dry basis. See Handbook Supplement 168-SF1. 2. 3 for conversions to wet basis.

Figure 3.10.4-3b. Percent of excess air from percent measured oxygen or carbon dioxide for various fuels. (*Courtesy North American Mfg. Co., Combustion Handbook, Cleveland, OH.*)

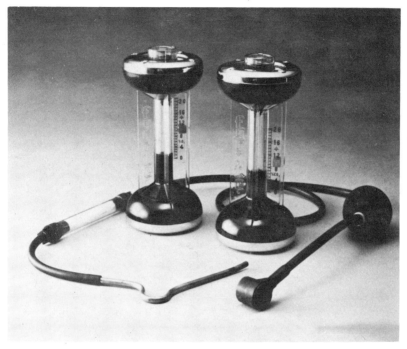

Figure 3.10.4-4. Oxygen and carbon dioxide measuring equipment. (*Courtesy Bacharach Instrument Co., Pittsburgh, PA.*)

measuring equipment can range from very sophisticated electronic equipment to simple absorption chemical apparatus. One of the simplest is shown in Figure 3.10.4-4. The results are read directly in percent of oxygen or carbon dioxide present.

3.11 MOIST AIR CONDITIONS

A psychrometric chart, shown in Figure 3.11-1, is very useful for finding the properties of moist air, and for determining the heat recovered or used in comfort applications. The dry bulb, or temperature measured by a thermometer, is plotted along the bottom scale. The amount of water vapor in the air, in grains per pound of dry air, is plotted along the vertical scale. The horizontal lines end on the left at the saturation curve; at these points the air is saturated, holding all the water vapor it can possibly hold. The curved lines from upper right to lower left are lines of constant relative humidity, or constant percent of water vapor content at saturation. Straight lines sloping from upper left to lower right are constant wet-bulb temperatures, measured

32 INDUSTRIAL AND COMMERCIAL HEAT RECOVERY SYSTEMS

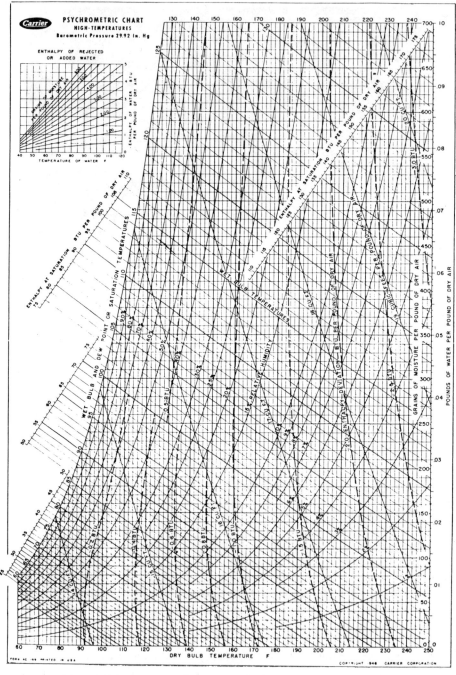

Figure 3.11-1. Psychrometric chart. (*Courtesy Carrier Corp., Syracuse, NY.*)

HEAT RECOVERY SURVEY 33

by a thermometer with a wet sleeve over the bulb held in a moving air stream, or moved in a sling psychrometer, or measured by a special instrument.

Dewpoint is an important factor in heat recovery. The heat recovery equipment should not cool the exhaust gas or air below its dewpoint, to avoid corrosion and water drainage problems. The intersection of dry bulb temperature with water vapor content, wet-bulb temperature or relative humidity is located on Figure 3.11-1. Moving horizontally to the left, the dewpoint is located at the intersection with the saturation curve.

The specific volume, in ft^3/lb of dry air, is found along the sharply sloping lines from upper left to lower right. The heat content of the air, enthalpy, in Btu/lb of dry air, is shown at the left of the saturation curve and is constant along lines of constant wet-bulb temperature. The amount of heat lost or recovered by air passing through heat recovery equipment is the weight flow, W, times the enthalpy difference of the air between the entering and leaving conditions:

$$H = W(h_l - h_e). \qquad (3\text{-}10)$$

Equation 3-10 is used for heating and cooling comfort calcuations where total heat loads must be considered. Equation 2-4 is used for process or heating calculations when only sensible heat loads are present. Latent heat is the heat transferred by virtue of the heat contained in the moisture transferred between the exhaust and supply. Sensible heat is the heat transferred by virtue of temperature change of the exhaust or supply. The total heat transferred is the sum of the latent and sensible heats.

3.12 MATCHING OF SOURCE AND USE FOR RECOVERED HEAT

As explained in Chapter 1, a heat recovery system has three components: a source of recoverable heat, a use for the recovered heat, and heat recovery equipment to transfer the heat from the source to the use. The preceding heat recovery survey has located the sources and uses, and succeeding chapters describe various types of heat recovery equipment. It remains now to combine the sources and uses, and to choose the heat recovery equipment, to form a successful heat recovery system. The choice of heat recovery equipment cannot be made until the sources and uses are matched into a heat recovery system, since the equipment choice depends upon the nature of the source and use. The matching of sources and uses into a heat recovery system is a trial-and-error process, based on certain guidelines:

1. Operating Cycle. The annual operating time should be as long as possible to secure the largest annual savings. The source and use should have similar operating cycles, and be in use during the same time period.

2. *Temperature of Source Exhaust.* The temperature of the source exhaust should be useful. It need not be above the use temperature, but should be sufficiently above the use supply temperature to allow operation of the heat recovery equipment.

3. *Flow Rates.* The flow of the source exhaust and use supply should be large enough to allow the use of heat recovery equipment and provide sufficient recoverable heat to be practical. Using equation 2-3, known and measured data, and estimating the amount of heat recovered, from Chapter 5, an estimate may be made of the practicality of available flow rates and temperatures. Very often, several sources or uses can be merged into the heat recovery system to provide enough flow and temperature to make the system practical. The average temperature, T_{avg}, resulting from the blending of streams 1, 2, . . . of *standard* volume flow, V, and temperature, T, is given by Equation 3-11:

$$T_{avg} = \frac{V_1 T_1 + V_2 T_2 + \cdots}{V_1 + V_2 + \cdots} \qquad (3\text{-}11)$$

4. *Balance of Heat.* An estimate of heat available and recovered may be calculated using measured flow rates, temperature, and the data presented in Chapter 5. Should this calculation show that the proposed pair of source and use either uses a small portion of the recoverable heat, or that the amount of heat recoverable is far less than the amount needed by the use, then a different pair should be tried. The first case is obviously a waste of recovered heat; the second forecloses the possibility of finding another source or blend of sources to supply the amount of heat needed by the use.

5. *Contaminants.* The source and use paired as a heat recovery system should be chosen, if possible, to reduce contamination. The heat recovery survey may have located sources containing contaminants too difficult to handle in the heat recovery equipment. The possibility exists when systems are blended that the contaminant problem will be mitigated by the dilution of the contaminant. This can happen when the dewpoint of moist air is lowered by reducing the relative humidity, or the concentration of a corrosive liquid is reduced by dilution. Conversely, it is also possible to worsen the contamination problem. A moist air stream mixing with a particulate-laden stream could produce a wet caking deposition problem. A cold air stream blending with a hot exhaust stream might produce condensation. Certainly, the choice of source and use for a heat recovery system that worsens the contamination problem is to be avoided.

6. *Proximity.* The distance between the matched source and use should be kept to a minimum. Extended runs of piping or ducting are expensive, lose heat, and require maintenance. Where runs of several hundred feet or more

are required, piping is the preferred method of heat transmission. Pipe is less expensive to insulate, easier to maintain, and less costly to install. Also, obstructions are more easily accommodated. The maximum run practical for a project depends on the cost savings, and must be determined for each heat recovery system.

7. *Installation*. The choice of a pair of source and use depends also on the physical arrangement of the source and use. There should be clear access to the area where the heat recovery equipment is to be installed. Any door or height limitations will have a bearing on equipment size. Obstructions between the source and use will increase the cost and complexity of the installation. This is particularly true of intervening conduit, piping, ducts, equipment, buildings, open spaces, etc. Similarly, structural limitations at the site may limit equipment weight or drainage. Adequate electrical services should be available. Finally, maintenance accessibility should be provided. Walkways, room to pull heat exchanger tubes, clearance for cranes, hoists where needed, are areas of consideration.

8. *Heat Recovery Equipment*. The selection of heat recovery equipment completes the preliminary definition of the heat recovery system. The nature of the source and the use dictates the type of heat recovery equipment. Various types of heat recovery equipment are described in Chapter 4, allowing a preliminary selection of a suitable type. Chapter 5 details methods of calculating the performance of many, but not all, types of heat recovery equipment. If these calculations show that the heat recovery system preliminary choice is feasible, then the financial evaluation of Chapter 7 can be carried out. If not, then a new pair of source and use should be tried. Chapter 6 provides examples of successful heat recovery systems.

REFERENCES

1. *Industrial Ventilation*, Committee on Industrial Ventilation, Lansing, MI.

4
HEAT RECOVERY EQUIPMENT

Heat recovery systems use heat recovery equipment to accomplish the transfer of heat from source to use. Equipment is available to handle gases, liquids, and vapors, or combinations of them. Specialized types have been developed to handle difficult requirements with little attention. This chapter presents descriptions of the various types of heat recovery equipment available. No attempt has been made to cover every manufacturer of every type, but rather to present a representative sampling of heat recovery equipment. Methods of performance calculation are presented in the next chapter, allowing calculation of system heat recovery once the heat recovery equipment has been chosen from the data in this chapter.

4.1 FUNDAMENTALS OF HEAT EXCHANGERS

Heat recovery equipment transfers heat from the source exhaust to the use supply using some form of heat exchanger. A brief review of heat transfer fundamentals and their use in calculating heat exchanger performance follows as a background in working with heat recovery equipment performance equations.

Heat is transferred between two regions when a temperature difference, ΔT, exists between them. There are three types of heat transfer: conduction, convection, and radiation. The transfer of heat across the temperature difference, ΔT, can occur by any one or any combination of the three types of heat transfer, depending on the conditions present.

Conduction is the transfer of heat within a material or between materials in close thermal contact. The amount of heat, H, transferred by a material having a thermal conductivity, k, a thickness, L, through an area A, is given by Equation 4-1:

$$H = \frac{kA \, \Delta T}{L}. \qquad (4\text{-}1)$$

If thermal conductivity is in units of Btu/hr/°F/ft²/ft, L is in ft, A is in ft², and ΔT is in °F, the units of H are Btu/hr. Values of k for most materials are listed in many handbooks.

Convection is the transfer of heat by mixing within a liquid or a gas, between areas having a temperature difference ΔT. Convection is particularly important close to the surface of a heat exchanger, where it controls the rate of heat transfer. A thin film of slow-moving gas or liquid exists next to the heat transfer surface. This film loses its temperature difference to the heat transfer surface. It is constantly mixed with gas or liquid still at the ΔT to maintain the heat transfer rate. The heat transfer, H, by convection, at a rate specified by the heat transfer coefficient, h, across the area, A, is given by Equation 4-2:

$$H = hA \, \Delta T. \qquad (4\text{-}2)$$

If the heat transfer coefficient, h, is in Btu/hr/°F/ft², A is in ft², and ΔT is in °F, the units of H are Btu/hr. Calculation of the value of h for various heat exchanger configurations may be found in the extensive available literature on the subject. Fortunately, for the purpose of heat recovery equipment performance calculations, manufacturers present heat transfer data making the calculation of h unnecessary.

Radiation is the transfer of heat by radiant energy from a heated surface or volume to a cold receiver. Radiation occurs only at elevated temperatures, when heat can be felt when facing the heated surface. Typically, radiant heat transfer becomes important at temperatures of 1200°F and above. Radiant energy is emitted by solids and tri-atomic gases, as carbon dioxide and water vapor. The heat transfer rate, H, by radiation, is governed by the emissivity, ϵ, a factor relating to surface properties, the area, A, a constant, F, relating to view of the receiver from the emitter, and the difference of the fourth powers of the emitter and receiver temperatures:

$$H = (0.173)(10^{-8}) \, \epsilon \, F(T_{\text{emitter}}^4 - T_{\text{receiver}}^4). \qquad (4\text{-}3)$$

The factor, $(0.173)(10^{-8})$ is the Stefan-Boltzmann constant, T is °F + 460, and H is in Btu/hr.

The overall heat transfer rate of a heat exchanger is specified by an overall heat transfer co-efficient, U. This coefficient, when no radiation is present, is determined by the thermal conductivity, k, and the convective heat transfer coefficient, h, using Equation 4-4:

$$\frac{1}{U} = \frac{1}{k} + \frac{1}{h}. \qquad (4\text{-}4)$$

The overall heat transfer rate of a heat exchanger with a heat transfer co-efficient, U, a heat transfer surface, A, and a temperature difference, ΔT, between the hot and cold streams is given by Equation 4-5:

$$H = UA\,\Delta T. \qquad (4\text{-}5)$$

The units are as before, with U in Btu/hr/°F/ft².

A number of coefficients, h, are involved in a practical heat exchanger. Each surface has its own coefficient, forming a series-type of process. The lowest value of the coefficient, h, controls the heat transfer rate, U, through Equations 4-4 and 4-5. Heat exchangers are designed to overcome low values of h by using large amounts of area, A. This is seen, for example, in a heat exchanger with gas as one of the materials. The heat transfer coefficient at the heat transfer surface, or film coefficient, h, for a gas is very low. These heat exchangers are built with fins or multiple plates to obtain as much surface as possible. Liquid film coefficients are much higher than gas film coefficients, reducing the need for large amounts of surface area in liquid heat exchangers.

In its most elementary form, a heat exchanger is a box with two flow passages separated by a surface. Only heat flows through the surface. Openings are provided in the box for flow through both passages. Three possible arrangements of flows are shown in Figure 4.1-1.

In a parallel-flow heat exchanger, both streams flow in the same direction. In a counter-flow heat exchanger, the streams flow in opposite directions. In a cross-flow heat exchanger, one stream flows at right angles to the other stream. At the entrance to the parallel-flow heat exchanger, the two streams are at their maximum temperature difference, ΔT_e. They leave with the minimum temperature difference, ΔT_l, because of heat transfer along the length of the heat exchanger. The temperature profile of the two streams along the heat exchanger length is shown in Figure 4.1-2a. In a counter-flow heat exchanger, the hot and cold streams enter from opposite ends of the heat exchanger and flow past each other in opposite directions. The temperature difference between the two streams is approximately constant along the heat exchanger length, as shown in Figure 4.1-2b.

The cross-flow heat exchanger temperature profile varies both through the heat exchanger and across the heat exchanger, because of heat transfer

HEAT RECOVERY EQUIPMENT

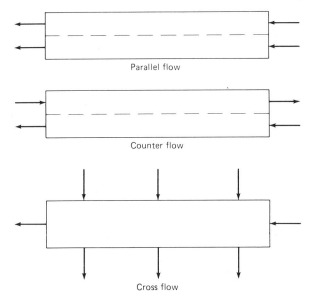

Figure 4.1-1. Flow patterns in a heat exchanger.

between the two streams both along and across the flow paths. The analysis is more complex than can be presented here. In practice, the cross-flow exchanger is handled by using correction factors with the equations developed in this section. The counter-flow heat exchanger will produce the largest heat transfer rate of the three types of flow patterns, because over the length of the heat exchanger it has the largest average ΔT, called ΔT_M, between the hot and

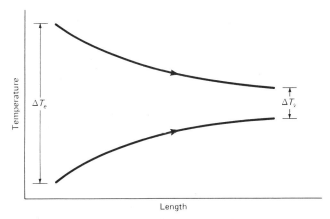

Figure 4.1-2a. Temperature profile in a parallel-flow heat exchanger.

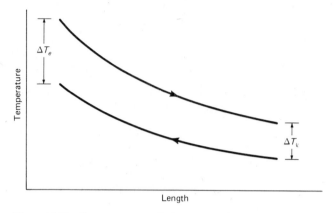

Figure 4.1-2b. Temperature profile in a counter-flow heat exchanger.

cold streams. Using ΔT_M to denote the average or "mean" temperature difference along the heat exchanger length, Equation 4-5, is written:

$$H = UA\ \Delta T_M. \qquad (4\text{-}6)$$

Equation 4-6 assumes that U remains constant along the heat exchanger, usually the case with a heat exchanger. If the further assumption is made that there is no heat lost to the surroundings and that heat is not conducted along the heat exchanger, ΔT_M is derived elsewhere (Reference 1), as:

$$\Delta T_M = \frac{\Delta T_1 - \Delta T_2}{2.3\ \log(\Delta T_1/\Delta T_2)}. \qquad (4\text{-}7)$$

ΔT_M is called the "log mean differential." The largest temperature difference at either end of the heat exchanger is used as ΔT_1, and the smallest temperature difference at the other end of the heat exchanger is used as ΔT_2. The "log" refers to the common logarithm.

The use of Equation 2-4 to calculate the amount of heat recovered by heat recovery equipment involves the use of ΔT, the temperature change through the heat recovery equipment. This ΔT is different from ΔT_M and cannot be used in its place. ΔT_M is a mean temperature difference inside the heat exchanger. ΔT is the actual change in temperature of a stream passing through the heat recovery equipment. Some heat recovery equipment manufacturers provide performance charts to calculate the value of ΔT based on flow conditions and equipment construction. The charts relate flow conditions to equipment "effectiveness," E. The effectiveness is defined as:

$$E = \frac{100\,\Delta T}{T_{ee} - T_{se}} \quad \text{or} = \frac{100\,\Delta T}{T_{se} - T_{ee}}. \tag{4-8}$$

Solving for ΔT:

$$\Delta T = \frac{E(T_{ee} - T_{se})}{100} \quad \text{or} = \frac{E(T_{se} - T_{ee})}{100} \tag{4-9}$$

Use the order of subtraction to produce a positive value in Equation 4-8 or 4-9. When the weight flows of the supply and exhaust streams are equal, Equation 4-9 is used for both ΔT_e and ΔT_s. When the flows are unequal, correction factors are supplied by manufacturers and are given in Chapter 5.

The foregoing discussion has dealt with heat exchangers with a stationery heat transfer surface and moving streams of gas or liquid. The name "recuperator" is applied to these heat exchangers. Instead of transferring heat across a surface, the heat can be brought to the gas or liquid in a "regenerator." A bed or porous material alternately contacts the hot and cold streams passing on its own heat to the cold stream or absorbing heat from the hot stream.

The rotary regenerator, or "heat wheel," scheme of operation is shown in Figure 4.1-3. A slowly turning wheel, packed with a porous medium, transfers heat from the exhaust to the supply. As it turns, the wheel is alternately heated and cooled.

The operating schematic of a stationery regenerator is shown in Figure 4.1-4. The stationery regenerator consists of several tanks filled with a heat-retaining packing. Three-way diverter valves before and after the tanks control the flow of exhaust and supply through the tanks. In scheme (b), exhaust flows through the left-hand tank, heating the packing material. Supply flows through the right-hand tank, and is heated while the packing material cools. The flow directions are reversed when the diverter valve positions are reversed. The stream temperatures oscillate between operating

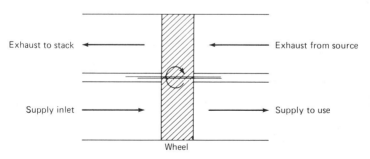

Figure 4.1-3. Rotary regenerator.

42 INDUSTRIAL AND COMMERCIAL HEAT RECOVERY SYSTEMS

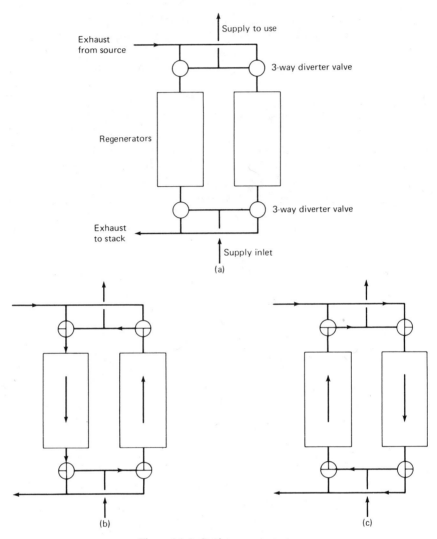

Figure 4.1-4. Stationery regenerator.

temperature limits for the stationery regenerator, and remain constant for the rotary regenerator.

4.2 GAS/GAS CURVED PLATE COUNTER-FLOW HEAT EXCHANGERS

The gas/gas curved plate counter-flow heat exchanger has thin corrugated metal plates as heat exchange surfaces. They are spaced about 0.5 in. apart.

HEAT RECOVERY EQUIPMENT 43

The plate edges are sealed with a plastic resin for use up to 400° F, or with a refractory cement for service to 1500° F. The resin or cement also bonds the heat exchanger core to the casing. The edge seal prevents cross-flow between the passages. At the ends, the plates are split in the center with a divider plate, and pairs of plate edges are joined. The resulting counter-flow gas flow is shown in Figure 4.2-1. The inlet flows are shown in black and the outlet flows are shown in white. As the flows pass through the heat exchanger, they are separated by the heat exchange surfaces. They emerge in the opposite half from where they entered. The corrugations provide mixing in the flow passages as well as stiffening the heat exchange surfaces. A typical unit is shown in Figure 4.2-2. These heat exchangers provide a range of effectiveness from 65% to 85%, depending on the flow rates and size. Pressure drop decreases with increasing effectiveness, from 0.2 in. water to 1.9 in. water. Sizes vary from openings of 6 in. by 24 in. to 48 in. by 48 in. and from 6 ft to 10 ft in length. The maximum flow rate is 8000 cfm. Units are connected in parallel to handle larger flow rates. For temperatures to 600° F, they are supplied with aluminum heat transfer surfaces and aluminum or galvanized steel enclosures, and are limited to a 2-in. water pressure differential across the surfaces. Steel and stainless steel materials are used up to 1500° F, allowing a 4-in. water pressure differential across the surfaces.

Figure 4.2-1. Curved plate counter-flow heat exchange flow path. (*Courtesy United Air Specialists, Inc., Cincinnati, OH.*)

44 INDUSTRIAL AND COMMERCIAL HEAT RECOVERY SYSTEMS

Figure 4.2-2. Curved plate counter-flow heat exchanger. (*Courtesy United Air Specialists, Inc., Cincinnati, OH.*)

The precautions of the previous chapter concerning particulate, moisture, condensation, and corrosion apply to curved plate counter-flow heat exchangers. Condensation should be avoided in the heat exchanger, since the condensate can be corrosive. Frosting and plugging of the exhaust flow passages will occur at temperatures close to freezing. Both these conditions are avoided by equipping the heat exchanger with face-and-bypass dampers. One set of dampers is located in the heat exchanger supply inlet. A second set is located to allow supply gas to be bypassed around the heat exchanger. A thermostat in the exhaust gas leaving the heat exchanger controls the two sets of dampers in opposite directions. As the face damper closes, the bypass damper opens, or the reverse occurs. The exhaust gas temperature reduction in the heat exchanger is limited by reducing the supply flow and increasing the bypass flow, resulting in less cooling of the exhaust gas.

A waterwash system is available where contamination cannot be filtered or avoided. Perforated spray tubes are provided at the top of each flow passage. The spray tubes are connected to a common manifold at one end of the heat exchanger. The manifold is supplied with hot water and detergent from a timed pumping system. The spray of hot detergent removes oils, greases, and soluble or loose particulate build-up. The interval between spray cycles and

HEAT RECOVERY EQUIPMENT 45

the length of washing are based on experience. Provisions are necessary to drain the wash solution and to bypass the heat exchanger during the wash cycle.

The heat exchanger rests on curbs when roof-mounted. It can be suspended from ceiling trusses by hanger rods attached to a frame under the exchanger. Inlet ducting should be designed to provide an even inlet velocity distribution. Sharp turns without turning vanes should be avoided; all transitions should be gradual. Otherwise, the flow will be unevenly distributed within the flow passages and the effectiveness will suffer. Both supply and exhaust fans either draw or push gas through the heat exchanger. If not, an excessive pressure differential is produced across the heat exchange surfaces. The supply is filtered to prevent dust accumulation in the inlet passages.

The curved plate counter-flow heat exchanger is available in many forms. Figure 4.2-3 shows a unit furnished with face-and-bypass dampers on the supply inlet. The small damper is the bypass damper.

Figure 4.2-4 shows a complete package air make-up unit for rooftop

Figure 4.2-3. Curved plate counter-flow heat exchanger equipped with face-and-bypass dampers. (*Courtesy United Air Specialists, Inc., Cincinnati, OH.*)

46 INDUSTRIAL AND COMMERCIAL HEAT RECOVERY SYSTEMS

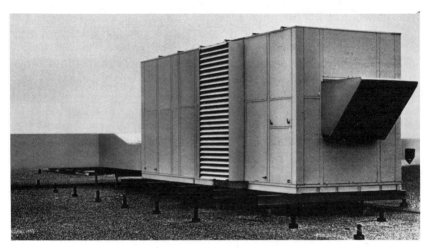

Figure 4.2-4. Complete package air make-up unit. (*Courtesy United Air Specialists, Inc., Cincinnati, OH.*)

mounting on curbs. It contains fans, filters, dampers, coils, controls, and motor starters in addition to the heat exchangers.

4.3 GAS/GAS FLAT PLATE COUNTER-FLOW HEAT EXCHANGERS

The gas/gas flat plate counter-flow heat exchanger contains many parallel flat plates in an enclosure. The plate spacing is maintained by a mechanical method. The ends are sealed to prevent cross-flow. Exhaust and supply gases flow in opposite directions in alternating flow paths. Means are provided to separate the exhaust and supply flows at the heat exchanger inlet and outlet. Some method of producing mixing in the flow passages is often provided. Because a wide variety of these heat exchangers are available, a representative example of each type is described in this section.

One type of gas/gas flat plate heat exchanger uses a thin metallic sheet as the heat transfer surface. The sheet is folded or pleated to provide flow passages 0.18 in. wide. Ribs and dimples pressed into the sheet maintain plate spacing and produce mixing in the flow passages. The method of construction is shown in Figure 4.3-1. The ends of the folded plates are sealed in resin or cement, as described in Section 4.2. The folded flat plates are mounted in an enclosure, to produce the flow pattern shown in Figure 4.3-2.

The supply gas flows in the opposite direction and on the opposite side of the heat transfer surface from the exhaust gas. Mixing of the two gas streams is prevented by the folds of the continuous heat transfer surface and the end

HEAT RECOVERY EQUIPMENT 47

Figure 4.3-1. Construction of folded flat plate counter-flow heat exchanger. (*Courtesy Des Champs Laboratories, Inc., East Hanover, NJ.*)

seals. The center panels on both sides cover the flow passage and can be removed for cleaning. A folded flat plate counter-flow heat exchanger is shown in Figure 4.3-3. These heat exchangers range in effectiveness from 62% to 74%, depending on flow. The pressure drop through them varies from 0.3 in. water to 1.4 in. water. The lower pressure drop is associated with the higher effectiveness. The heat exchangers are available as standard modules rated from 500 to 10,000 cfm. Higher flow rates are handled by ducting the required number of modules in parallel. The pressure difference across the plate is limited to 10 in. water at 450° F for standard aluminum construction. For

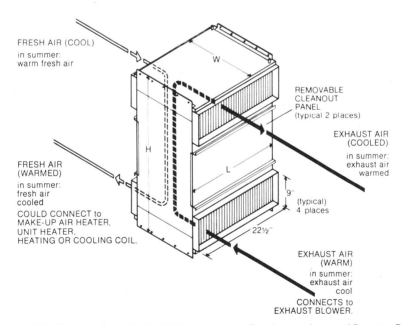

Figure 4.3-2. Flow pattern of folded flat plate counter-flow heat exchanger. (*Courtesy Des Champs Laboratories, Inc., East Hanover, NJ.*)

48 INDUSTRIAL AND COMMERCIAL HEAT RECOVERY SYSTEMS

Figure 4.3-3. Folded flat plate counter-flow heat exchanger. (*Courtesy Des Champs Laboratories, Inc., East Hanover, NJ.*)

temperatures to 450° F, aluminum, galvanized steel, or stainless steel housings are used, the latter for corrosive applications. Up to 1000° F, stainless steel plates with aluminized steel or stainless steel enclosures are available. All stainless steel construction should be used between 1000° F and 1400° F.

Precautions concerning dust, particulate, corrosion, and condensation given in Section 4.2 apply equally to these heat exchangers. Special coatings are available to permit less costly materials to be used in corrosive situations. Epoxy coatings are available on aluminum for mildly corrosive applications, such as swimming pools. The epoxy coating temperature limitation is 150° F. Baked phenolic coatings are available for use to 450° F. A waterwash system may be used to remove greases, oils, and non-hardening particulates. A detergent-water solution is sprayed into the heat exchanger by a traversing spray nozzle on the exhaust side. A drain at the bottom of the heat exchanger removes the wash water. The spray system is actuated by a timer. The wash cycle lasts from 10 to 25 minutes, depending on the equipment and the extent of the deposition. During the wash cycle, the heat exchanger is bypassed or shut down. To prevent frosting, face and bypass dampers are available, as described in Section 4.2.

The manufacturer offers an interesting and useful variation of normal heat recovery. Clean water is sprayed into the exhaust inlet of the heat exchanger used for air makeup. The water evaporates and cools the exhaust air by an amount depending on the exhaust relative humidity and air temperature. The cooled exhaust air removes heat from the incoming supply air, without humidifying it. The cost savings from pre-cooling the supply air can be considerable. The wet side of the heat exchanger must be epoxy coated and sealed with mastic to avoid corrosion. No pump is required; the spray nozzles operate from water main pressure. Figure 4.3-4 shows a package rooftop unit. Exhaust and supply fans are arranged to pull air through the heat exchangers, avoiding unnecessary pressure differential across the heat exchanger plates. A separate bypass inlet with damper is provided for partial flow conditions. The heat exchanger modules are arranged to operate in parallel. The outside air entrance is filtered, as is the bypass air. Supplemental coils are provided for additional heating or cooling. The unit is normally mounted on a curb with direct connections to exhaust and supply ducts. Simpler package units are available for industrial applications, without dampers, filters, and coils. They are roof-mounted, suspended from ceiling trusses, or made part of the process.

A variation of the flat plate counter-flow heat exchanger is shown in Figure 4.3-5. The heat exchanger construction is shown in Figure 4.3-6.

Alternate ends of the folded heat transfer surfaces are mechanically fastened or welded together, instead of being sealed in a resin or cement. The exhaust gas flows directly through the heat exchanger, reducing the obstruc-

Figure 4.3-4. Package rooftop flat plate counter-flow heat exchanger air makeup unit. (*Courtesy Des Champs Laboratories, Inc., East Hanover, NJ.*)

50 INDUSTRIAL AND COMMERCIAL HEAT RECOVERY SYSTEMS

Figure 4.3-5. Flat plate counter-flow heat exchanger. (*Courtesy Exothermics Energy Recovery Systems, Toledo, OH.*)

tion to the exhaust flow. The supply flow still follows a U-shaped path through the heat exchanger. The plates are spaced 0.5 in. apart, with formed projections to maintain the spacing. The combination of straight-through flow and wide plate spacing provides good access for cleaning the exhaust flow passages. Side panels are removable. The materials of construction, performance, and other details are similar to those previously described. This same type of construction is also used in process heat exchangers for metallurgical applications. The heat transfer surfaces are made from flat plates, heavy enough to withstand operating pressure differentials without obstructing the exhaust gas passages. Mechanical methods are used to clean the exhaust gas passages.

4.4 GAS/GAS CROSS-FLOW HEAT EXCHANGERS

Cross-flow heat exchangers are made with parallel flat heat transfer surfaces spaced by fins, rods, or projections formed in the plates. Exhaust and supply

HEAT RECOVERY EQUIPMENT 51

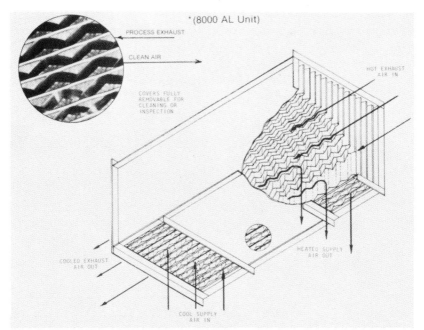

Figure 4.3-6. Construction of flat plate counter-flow heat exchanger. (*Courtesy Exothermics Energy Recovery Systems, Toledo, OH.*)

gases flow in alternate passages between the plates. The spacers aid the heat transfer process by picking up additional heat from the exhaust gas, conducting it to the supply gas passages, and transferring it to the supply gas. Several different constructions are described in this section. A cross-flow heat exchanger is shown in Figure 4.4-1.

The heat exchanger is available as a single unit, 50 in. deep in each flow direction and from 14 in. to 154 in. in height, depending on the amount of flow. The flat heat transfer surfaces are separated by 5/32-in. rods lying in the direction of flow, spaced 6 in. apart. The rod direction alternates in adjacent flow passages to accommodate the cross-flow design. The rods extend the length of the flow path. The ends of the plates are formed on two sides to form a press-fit seal in the non-flow direction. The plate assembly is insulated from the frame to prevent heat loss and to allow for thermal expansion. Punched flanges are provided on the frame for duct connections. The heat exchanger is available in aluminum, aluminized steel, and stainless steel, and will operate at temperatures up to 1500° F. The pressure difference across the plate is limited to 6 in. water. As the effectiveness falls, the pressure drop rises. At an effectiveness of 60%, the pressure drop is 0.23 in. water. At an effectiveness of 50%, the pressure drop is 0.65 in. water. Units can be staged to produce a higher effectiveness. Auxiliary face-and-bypass dampers, water wash systems,

52 INDUSTRIAL AND COMMERCIAL HEAT RECOVERY SYSTEMS

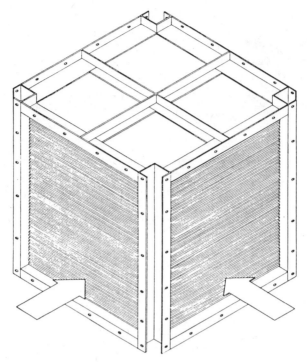

Figure 4.4-1. Cross-flow heat exchanger. (*Courtesy National-Standard Machinery Systems Division, Rome, NY.*)

and fan and filter assemblies are available. Precautions concerning contaminants, cleaning methods, frost prevention, etc., previously stated, should also be observed with these cross-flow heat exchangers.

A cross-flow heat exchanger is available from treated paper. Parallel heat transfer surfaces of treated paper are separated by folded paper spacers, as shown in Figure 4.4-2.

The heat exchanger is formed into a block and becomes part of a package heat exchanger unit. The paper heat transfer surfaces transmit both sensible and latent heat. The manufacturer claims the thermal conductivity of the paper to be within 3% of copper and aluminum, transferring sensible heat. The paper transfers moisture from the more humid flow to the less humid flow, resulting in the transfer of latent heat. The combination of the two modes of heat transfer results in greatly increased summer heat recovery for building ventilation applications, and moderate increases for winter heat transfer conditions. The heat exchanger is intended for air ventilation applications where dust loading is removed by filters, and temperatures are within the range of normal ventilation conditions. The heat transfer surfaces

HEAT RECOVERY EQUIPMENT 53

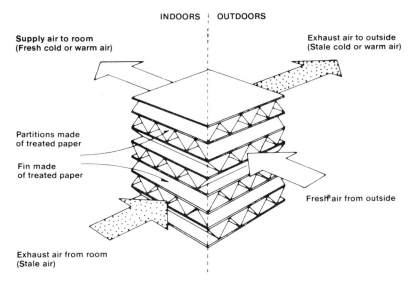

Figure 4.4-2. Construction of paper cross-flow heat exchanger. (*Courtesy Mitsubishi Electric Sales America, Inc., Compton, CA.*)

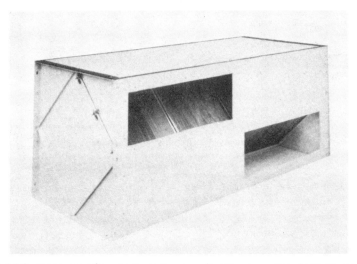

Figure 4.4-3. Paper cross-flow heat exchanger unit. (*Courtesy Mitsubishi Electric Sales America, Inc., Compton, CA.*)

54 INDUSTRIAL AND COMMERCIAL HEAT RECOVERY SYSTEMS

are spaced 0.1 in. to 0.2 in. apart. Pressure drops are up to 2 in. water. Sensible and latent heat effectivenesses are up to 85%. A package heat exchanger is shown in Figure 4.4-3.

The element is placed on a diagonal in the enclosure, providing for straight-through ducting. A typical installation is shown in Figure 4.4-4.

In this scheme, the intake is supplied with a pre-filter. In case of frosting problems, a face-and-bypass damper is included. The pressure difference across the heat transfer surfaces is minimized by having both exhaust and supply fans either draw or push air through the heat exchanger. At a pressure difference of 2 in. water, 10.5% leakage occurs; at 4 in. water, a 13.0% leakage occurs. Leakages of various contaminants are shown in Figure 4.4-5, where "S" indicates the standard flow.

4.5 GAS/GAS SHELL-AND-TUBE HEAT EXCHANGERS

Shell-and-tube heat exchangers are not frequently used in gas/gas service because they are more expensive to construct and are bulkier than other types. In dirty applications, with high particulate or high condensible vapor concentration, shell-and-tube heat exchangers are used because they can be easily cleaned. Tube replacement is not difficult, making these heat exchangers especially useful in corrosive applications. A schematic diagram of a shell-and-tube heat exchanger is shown in Figure 4.5-1. Shown here is a

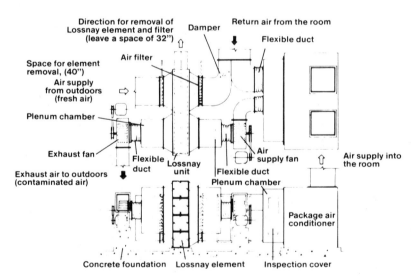

Figure 4.4-4. Typical ventilation application of paper cross-flow heat exchanger. (*Courtesy Mitsubishi Electric Sales America, Inc., Compton, CA.*)

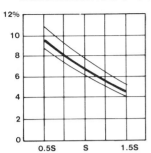
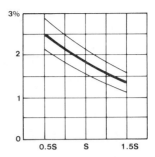
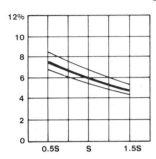
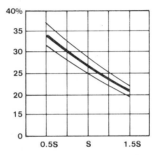

Figure 4.4-5. Leakage of various contaminants in paper cross-flow heat exchanger. (*Courtesy Mitsubishi Electric Sales America, Inc., Compton, CA.*)

two-pass design, meaning that the gas within the tube passes the length of the heat exchanger twice before leaving the heat exchanger. The exhaust gas enters at EI and passes through the upper half of the tubes to the end of the heat exchanger. The gas reverses direction in the end chamber and passes through the tubes a second time, before leaving at EO. Supply gas enters at FI, and passes outside the bottom half of the tubes around baffles. At the end of the heat exchanger, the supply gas reverses direction, flowing around the top tube and baffles. The supply gas leaves at FO. In both halves of the heat exchanger, counterflow operation is maintained. One end of the tube bundle is bolted to flanges of the heat exchanger shell. The other end slides freely inside the shell to allow for differences in thermal expansion between the shell and the tube bundle. A seal is provided to prevent mixing of the supply and exhaust flows. The tubes are rolled into the tube sheets. A view of the heat exchanger and the tube bundle is shown in Figure 4.5-2.

The heat exchanger has an effectiveness of 65%. Sizes are available from

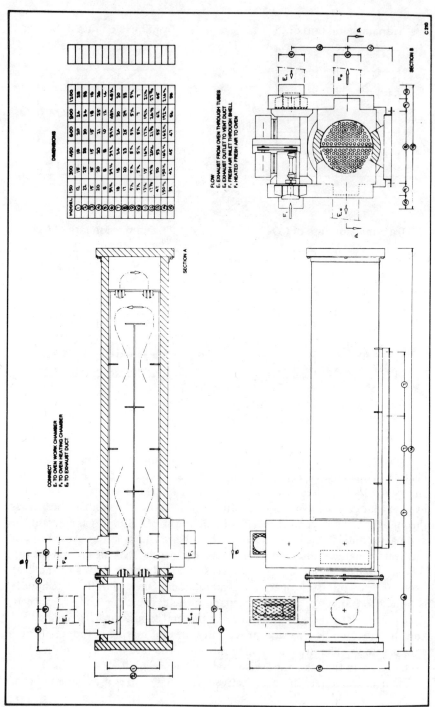

Figure 4.5-1. Schematic diagram of gas/gas shell-and-tube heat exchanger. (*Courtesy Bayco Industries of California, San Leandro, CA.*)

HEAT RECOVERY EQUIPMENT 57

Figure 4.5-2. View of complete sheet-and-tube heat exchanger, and tube bundle. (*Courtesy Bayco Industries of California, San Leandro, CA.*)

150 standard cfm to 1200 standard cfm. The pressure drop varies with size from 1 in. to 2 in. water. It is made from carbon steel, with 1-in. diameter tubes, 0.25-in.-thick tube sheets, and 3-in. insulation around the shell. The operating temperature limit is 650° F. The end pieces of the shell are easily removed, allowing cleaning of the tubes with brushes or scrapers. The unit shown in Figure 4.5-2 is furnished with blowers, motors, and pressure switches to warn against loss of flow.

Another shell-and-tube heat exchanger is shown in Figure 4.5-3. This is a single-pass heat exchanger. The gas enters at one end, passes through all the tubes, and leaves at the other end. The tubes are 18-gauge seamless steel, $1\frac{1}{2}$-in. diameter, and 20 ft long. The shell is a 14-gauge steel casing. The tubes are rolled into the header plates. Steel is used for applications up to 800° F, and stainless steel is used for those up to 1200° F.

Shell-and-tube heat exchangers have been used in baking and curing ovens, dirty or high particulate applications, and applications where easy accessibility is important.

4.6 GAS/GAS HEAT PIPE HEAT EXCHANGERS

A heat pipe is a gas/gas heat exchanger combining the processes of heat transfer, vaporization, and condensation. A heat pipe heat exchanger actually consists of numbers of heat pipe elements contained in a frame. A schematic of a heat pipe element is shown in Figure 4.6-1.

58 INDUSTRIAL AND COMMERCIAL HEAT RECOVERY SYSTEMS

Figure 4.5-3. Single-pass shell-and-tube heat exchanger. (*Courtesy United Air Specialists, Inc., Cincinnati, OH.*)

The heat pipe element consists of a hollow tube, sealed at both ends. The exterior of the tube is fitted with metallic fins to aid the transfer of heat from the gas to the tube. The interior of the tube has a fibrous wicking material around the tube wall and is filled with a working fluid, usually a fluorocarbon compound. The interior is evacuated to the working fluid vapor pressure.

When hot exhaust gas passes over one end of the heat pipe element, some of the liquid in the tube evaporates, driving up the vapor pressure in the hot end. The internal pressure excess over the cold supply end pushes some of the vapor to the cold end. Supply gas passing over the cold end removes heat from the fins, tube, and, eventually, the vapor, condensing it back to a liquid. The

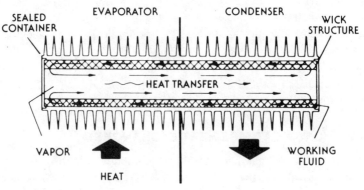

Figure 4.6-1. Schematic of a heat pipe element. (*Courtesy James Howden America, Newton, NJ.*)

liquid is transported by capillary action in the wicking structure back to the hot end. The exhaust air has been cooled, and the supply air warmed. The process repeats itself continuously, using no energy or moving parts.

The assembly of many heat pipes into a heat exchanger is shown in Figure 4.6-2. A dividing partition separates the exhaust and supply flows. The width and height of each side of the heat pipe exchangers are designed to provide an approach velocity of about 500 fpm for each flow rate with reasonable duct connection dimensions. The number of rows of heat pipe elements in the flow direction is selected to provide the required effectiveness and pressure drop; up to seven rows are generally supplied. Effectivenesses vary from 50% to 80%, averaging about 60%. Pressure drops range from 0.2 in. water to 2.0 in. water. Heat pipe exchangers are available in size from 1 ft in height by 2 ft in width to 5 ft in height by 20 ft in width.

Heat pipes are normally made from aluminum for service up to 525° F, copper for use up to 425° F, and carbon steel for service up to 800° F. For corrosive applications, the elements are made from copper, coated with a baked phenolic resin, or given a chromate-anodized finish. Fin spacings from 5 to 14 fins per in. are available. The wider fin spacing is used with contaminated gases. Heat pipes are available with wider fin spacing on the exhaust side than on the supply side, with consequent cost and material savings for the supply side.

Heat pipe manufacturers attach the fins to the heat pipe tubes in different ways. In one case, the pipe and fins are integral; the fin is extruded from the pipe wall. In this case, the manufacturer claims better joint thermal conduction and greater resistance to corrosion. Another manufacturer produces heat pipes by mechanically expanding the tube to lock tightly into the sheet metal fin; this construction allows the use of corrugated and rectangular fins, improving the performance of the heat pipe.

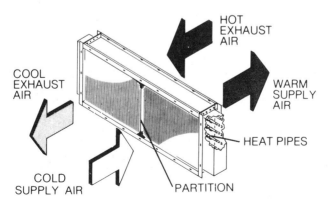

Figure 4.6-2. Heat pipe heat exchanger. (*Courtesy Q-dot Corp., Dallas, TX.*)

60 INDUSTRIAL AND COMMERCIAL HEAT RECOVERY SYSTEMS

Face-and-bypass dampers are also used with heat pipe heat exchangers to avoid frosting of the exhaust side. As the exhaust gas temperature in the heat pipe approaches the frosting point, the face damper on the supply side closes and the bypass damper on the same side opens. The amount of cooling in the heat exchanger is reduced, avoiding the possibility of frost in the exhaust side. A water wash system is also available, consisting of a built-in set of spray nozzles or a flooding flow from an overhead distribution tray. Spray wash systems are limited to four row widths. If matting fibers are present, a filter is recommended. The performance of a heat pipe can be improved by tilting the heat pipe slightly downward on the exhaust side, aiding the flow of condensed liquid through the wicking. Heat pipes are available with a variable tilt drive, controlled by the supply outlet temperature. The tilt is changed to allow for seasonal supply temperature changes or seasonal variations in flows of exhaust or supply. A typical heat pipe heat exchanger is shown in Figure 4.6-3.

The heat pipe unit is installed in the ducting system where exhaust and supply flows can be brought together. The supply and exhaust ducts are bolted to flanges on the rim and dividing partition, completely separating the flows. The heat exchanger can be mounted vertically, horizontally, or at an angle to the flow, but not tilted beyond the manufacturer's recommendation. Fans are arranged to push or draw air through the exchanger as there are no pressure limitations. The heat pipe is also available as part of package units for rooftop or suspended mountings. These units are complete, with fans, filters, dampers, electrical controls, and heating and cooling coils.

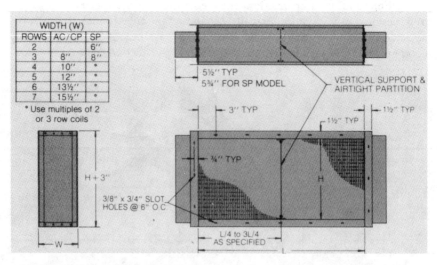

Figure 4.6-3. Typical heat pipe heat exchanger. (*Courtesy Q-dot Corp., Dallas, TX.*)

4.7 HEAT WHEELS

The heat wheel is a gas/gas rotary regenerative heat exchanger. A corrugated metal or fibrous material is wound into a round, narrow cylinder, and supported by a metal structure, to form a heat wheel. The center of the structure contains a bearing and shaft mechanism. The heat wheel is enclosed in a metal frame. The frame contains the drive motor and belt or gears, mounting surfaces for the ducts, a center flow divider, seals, and a purge section. A typical heat wheel is shown in Figure 4.7-1. These wheels can range in diameter from 3 ft to over 12 ft, with pressure drops from 0.1 in. water to 1.5 in. water.

Exhaust and supply ducts are connected to halves of the frame and the center flow divider. A drive motor slowly rotates the wheel between the exhaust and supply flows. In one side, the corrugated material is heated. In the other side, it is cooled. Since the wheel steadily turns, the heat transfer process is continuous.

Figure 4.7-2 illustrates the use of the heat wheel for air makeup and process

Figure 4.7-1. Typical heat wheel. (*Courtesy The Wing Co., Cranford, NJ.*)

use. The heat wheel element for summer makeup air use is coated with a hygroscopic salt for transferring moisture. This results in a gain of latent heat as well as sensible heat.

Two methods of sealing are used to prevent cross-flow between the exhaust and supply flows: seals and purging. They are illustrated in Figures 4.7-3 and 4.7-4.

Circumferential seals are provided around the wheel periphery to prevent leakage around the wheel. A seal is attached to the center flow divider to prevent flow between the exhaust and supply ducts. The heat wheel passages contain carry-over gas which is removed in the purge section. A portion of the supply flow is diverted through the purge section to flush the wheel passages, leaving with the exhaust gas. The amount of supply gas lost in the purge section is made up by over-sizing the supply fan. The amount of exhaust carry-over is held to less than 0.4% by volume.

A second method of purging provides flow passages as part of the central flow divider, on both sides of the heat wheel. An external blower continuously exhausts the flow passages, removing face seal leakage and purging gas. This type of purge is especially useful for heat wheels operating at elevated temperatures, where a flexible face seal cannot be used. The expected carry-over from this type of purging and face seal is less than 1%.

The heat wheels are driven at a fairly slow speed, on the order of 20 rpm. The speed is sometimes varied in response to the temperature of the supply gas, maintaining a desired supply outlet temperature. For low or moderate temperature applications, a drive belt surrounds the wheel and is driven by a pulley on the motor. For higher temperature applications, a chain drives a sprocket on the wheel shaft. The slow speeds involved in heat wheels result in low drive maintenance.

Heat wheels intended for operations up to 300° F are made with fibrous material or aluminum heat exchange media. For higher temperature use, various steel and stainless steels are available. At high temperatures, special care must be used in selecting seals, bearings, and drives.

The fibrous material used in moisture-transferring heat wheels is impregnated with a salt, often lithium chloride. The salt is a powerful hygroscopic agent, picking up moisture from the higher humidity gas passing through the wheel and transferring it to the lower humidity gas. In the summer, the exhaust air has been cooled and de-humidified by the air conditioning system. The supply air is hot and moist. The heat wheel element is cooled and dried by the conditioned exhaust air. It rotates into the supply air, and absorbs both heat and moisture from the supply air. The amount of heat removed from the supply air through cooling is the sensible energy saving. The amount of moisture removed by the heat wheel from the supply air does not have to be removed by the air conditioning system, and produces

HEAT RECOVERY EQUIPMENT 63

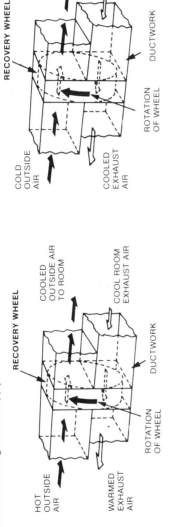
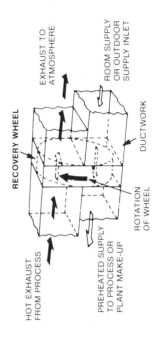

Figure 4.7-2. Heat wheel operation. (*Courtesy The Wing Co., Cranford, N.J.*)

64 INDUSTRIAL AND COMMERCIAL HEAT RECOVERY SYSTEMS

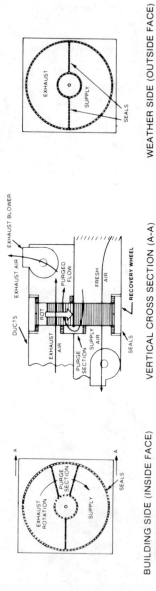

Figure 4.7-3. Methods of sealing and purging a heat wheel. (*Courtesy The Wing Co., Cranford, N.J.*)

HEAT RECOVERY EQUIPMENT 65

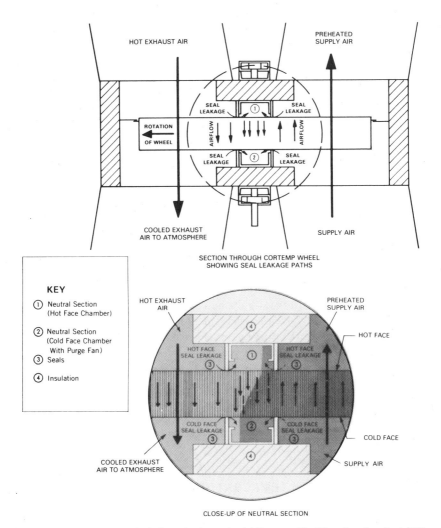

Figure 4.7-4. Alternate method of purging heat wheel. (*Courtesy The Wing Co., Cranford, NJ.*)

a latent energy saving. The additional latent energy saving produces an effectiveness approaching 90% for these heat wheels. Their sensible effectiveness alone approaches 80%. In winter operation, the moisture added by the heat wheel is at room temperature, saving the heat required to heat humidifier water to room temperature.

Heat wheels are suitable for operation on dust loadings that can largely be removed by filters. Where the contaminant does not adhere, heat wheels are

66 INDUSTRIAL AND COMMERCIAL HEAT RECOVERY SYSTEMS

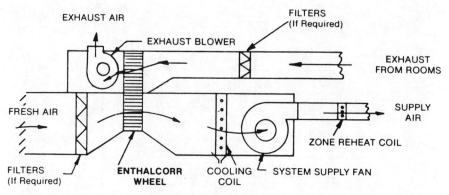

Figure 4.7-5. Heat wheel package unit schematic diagram. (*Courtesy The Wing Co., Cranford, NJ.*)

self-cleaning, when the fans are arranged as shown in Figure 4.7-5, a schematic of a package roof-top unit.

The supply and exhaust flows are in opposite directions. Any deposit of non-sticking material at the supply face is pulled out the exhaust outlet by the leaving exhaust gas. This fan arrangement also permits the purge section to operate, by removing the purge gases. Face dampers are fitted to the supply inlet and bypass dampers are placed between the exhaust and supply chambers. If frosting should occur in the exhaust due to excessive supply cooling, the face damper is partially closed and the bypass damper partially opened, reducing the amount of cooling. A coil is also shown in the supply chamber, providing heating or cooling. A coil can also be used in place of the face-and-bypass dampers to eliminate frosting by pre-heating the supply gas before the heat wheel.

4.8 GAS/GAS HIGH TEMPERATURE HEAT EXCHANGERS

The gas/gas heat exchangers described in previous sections depend upon convective processes to transfer heat from the gas to the heat transfer surfaces. Above 1400° F, radiation becomes the most important heat transfer process, requiring its own design considerations. The radiation emanates from tri-atomic gases, such as water vapor and carbon dioxide, and increases with the thickness of the gas layer being viewed by the heat transfer surface. Radiation heat transfer increases rapidly with temperature increases, and with the surface exposed to the radiating gases. In this section, two types of heat exchangers available as pre-engineered packages will be discussed. A number of other high-temperature heat exchangers available as custom-designed units will also be described.

HEAT RECOVERY EQUIPMENT 67

Figure 4.8-1. High-temperature cross-flow heat exchanger. (*Courtesy GTE Products Corp., Towanda, PA.*)

The cross-flow heat exchanger has been adapted to high temperatures by molding it from ceramics. The ceramic cross-flow heat exchanger is manufactured from cordierite, a magnesium aluminum silicate. It has a low coefficient of thermal expansion and good resistance to corrosion. The exhaust gas passages are vertical, and are 0.300 in. by 0.750 in. The supply gas passages are horizontal, and are 0.125 in. by 0.750 in. The typical heat exchanger package is shown in Figure 4.8-1.

One face of the enclosure is spring-loaded, to allow for thermal expansion. The interior of the enclosure contains a ceramic lining shaped to hold the cross-flow heat exchanger. The spring-loaded face seals the exhaust gas side, preventing mixing of the gases. These units have effectivenesses of 30% to 50%. The heat exchanger package is roughly a 15-in. cube. The heat exchanger block is shown in Figure 4.8-2.

Exhaust gas pressure drops ranges from 0.015 in. water to 1.0 in. water. Supply gas pressure drop varies from 0.15 in. water to above 10.0 in. water. Flow rates to 15,000 standard cfm and temperatures to 2600°F can be handled by this heat exchanger block.

68 INDUSTRIAL AND COMMERCIAL HEAT RECOVERY SYSTEMS

Figure 4.8-2. Heat exchanger block. (*Courtesy GTE Products Corp., Towanda, PA.*)

The cross-flow high-temperature heat exchanger is a particularly compact arrangement for moderate flow rates. It is easily connected into a system, tolerates moderate amounts of dirt, and is easily cleaned. It has been used both as a self-contained heat exchanger and as part of a packaged combustion air pre-heat burner.

The counter-flow radiant tube is another type of high-temperature heat exchanger, and is shown in Figure 4.8-3.

Exhaust gases enter the large central tube of the heat exchanger, passing upwards and out of the exchanger into the exhaust stack. Supply gas or air enters at the bottom of the exchanger; it passes upwards through a narrow annular channel, reverses direction at the top, and flows downwards through another narrow annular channel. The heated gas or air leaves at the bottom of the heat exchanger. The high-temperature gases in the center tube transfer heat mostly by radiation to the tube wall. The gas or air flowing next to the wall picks up heat from the wall by convection. Gas or air in the outer chamber helps to reduce external heat losses and increases the heat transfer surface. All

HEAT RECOVERY EQUIPMENT 69

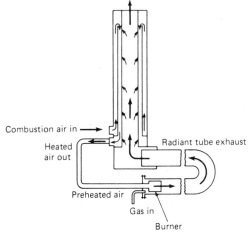

Figure 4.8-3. Counter-flow radiant tube heat exchanger. (*Courtesy Holcroft Div., Livonia, MI.*)

heat transfer is in the counter-flow direction, and the channels are designed for high heat transfer rates, sometimes with fins or other means of increasing heat transfer surface on the supply side.

The double supply gas or air passage is not used on all radiant tube heat exchangers. In some cases, the supply gas or air enters at the bottom of the heat exchanger and flows upwards through a single supply gas or air channel. The external surface is fully insulated. An example of this heat exchanger is shown in Figure 4.8-4.

These heat exchangers are generally manufactured from stainless steels, and operate with exhaust gas temperatures up to 2300° F. They have effectivenesses in the range of 40% to 50%. The supply pressure drop is from 1 in. to 11 in. water, depending on the flow. The exhaust pressure drop is supplied by the stack draft. They vary in length from 4 ft to 17 ft, and in diameter to 4 ft.

The radiant heat exchangers described above are suited for installing either on individual burners of a furnace, for pre-heating combustion air, or as a heat exchanger for process applications. The use of heat exchangers for pre-heating combustion air is described in the next section. For use in process applications, these heat exchangers should be installed with exhaust gas bypass ducting and safety controls to prevent overheating and consequent damage to the internal structure, in case of supply gas flow interruption.

There are other types of high temperature radiant heat exchangers available, which are described below. These are usually available as custom-designed units.

The heat exchanger shown in Figure 4.8-5 has a tubular structure shown in

70 INDUSTRIAL AND COMMERCIAL HEAT RECOVERY SYSTEMS

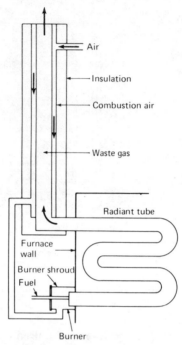

Figure 4.8-4. Counter-flow radiant tube heat exchanger. (*Courtesy Thermal Transfer Corp., Monroeville, PA.*)

Figure 4.8-6. The tubular structure is suspended freely inside a refractory lined casing. Supply gas or air enters at the bottom and leaves at the top, as shown in Figure 4.8-7.

The cold flow at the hottest exhaust temperature area maintains the tube temperature within operable limits. The same idea is used for certain high temperature radiant tube heat exchangers where the supply gas or air is brought in both at the top and bottom of the annular passages, and removed at the middle of the heat exchanger. The parallel-flow at the exhaust entrance prevents excessive metal temperature and the counter-flow at the top takes into account the convective heat transfer advantage of counter-flow design.

Cross-flow tubular heat exchangers are also available for high-temperature applications, as shown in Figure 4.8-8.

The exhaust gas flows through the duct and outside the tubes. The supply gas or air flows through the tubes. It flows into the upstream section and out of the downstream section to maintain permissible metal temperatures. Where these units are used on non-radiation applications, the supply gas or air direction is reversed to provide counter-flow heat transfer.

A stationary regenerative heat exchanger is available for high-temperature

HEAT RECOVERY EQUIPMENT 71

Figure 4.8-5. Tubular counter-flow high-temperature heat exchanger. (*Courtesy Thermal Transfer Corp., Monroeville, PA.*)

heat recovery. The flow path for this type of regenerator was shown in Figure 4.1-4, and an actual regenerator is shown in Figure 4.8-9.

The tanks are filled with non-packing stoneware, thermally stable to 2300°F. The stoneware has 70% free space, producing a low flow resistance. For high-temperature use, the flow is distributed evenly across the bed height and proceeds evenly across the bed width to a collection manifold, where it is gathered and exhausted. On the reverse cycle, the collection manifold serves as the intake distribution manifold. The system is shown in Figure 4.8-10.

The regenerator has an effectiveness of 85%, and is resistant to corrosive attack because of the stoneware internals. Typically, a unit for 6,000 cfm has tanks 2.5 ft diameter and 6.5 ft high. A unit for 24,000 cfm has tanks 7 ft high and 5 ft in diameter. The units are used at temperatures up to 1800°F, and sizes are available from 2000 cfm to 140,000 cfm and larger.

72 INDUSTRIAL AND COMMERCIAL HEAT RECOVERY SYSTEMS

Figure 4.8-6. Tubular heat exchange structure. (*Courtesy Thermal Transfer Corp., Monroeville, PA.*)

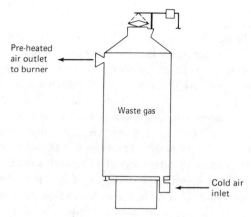

Figure 4.8-7. Flow path of tubular radiant tube heat exchanger. (*Courtesy Thermal Transfer Corp., Monroeville, PA.*)

HEAT RECOVERY EQUIPMENT 73

Figure 4.8-8. Cross-flow tubular high-temperature heat exchanger. (*Courtesy Thermal Transfer Corp., Monroeville, PA.*)

Figure 4.8-9. Stationery regenerative heat exchanger. (*Courtesy Regenerative Environmental Equipment Co., Inc., Morris Plains, NJ.*)

74 INDUSTRIAL AND COMMERCIAL HEAT RECOVERY SYSTEMS

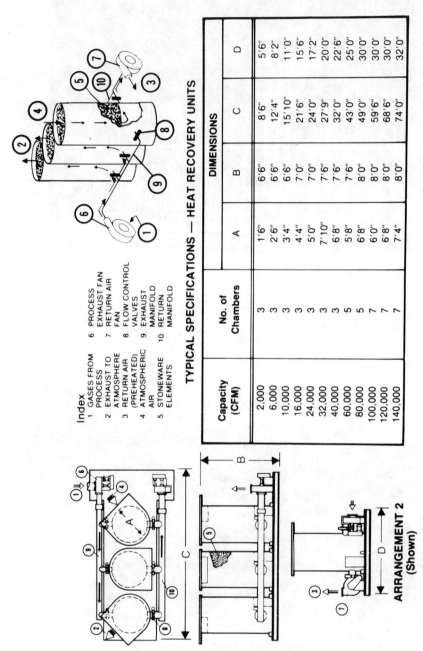

Index
1. GASES FROM PROCESS
2. EXHAUST TO ATMOSPHERE
3. RETURN AIR (PREHEATED)
4. ATMOSPHERIC AIR
5. STONEWARE ELEMENTS
6. PROCESS EXHAUST FAN
7. RETURN AIR FAN
8. FLOW CONTROL VALVES
9. EXHAUST MANIFOLD
10. RETURN MANIFOLD

TYPICAL SPECIFICATIONS — HEAT RECOVERY UNITS

Capacity (CFM)	No. of Chambers	DIMENSIONS			
		A	B	C	D
2,000	3	1'6"	6'6"	8'6"	5'6"
6,000	3	2'6"	6'6"	12'4"	8'2"
10,000	3	3'4"	6'6"	15'10"	11'0"
16,000	3	4'4"	7'0"	21'6"	15'6"
24,000	3	5'0"	7'0"	24'0"	17'2"
32,000	3	7'10"	7'6"	27'9"	20'0"
40,000	3	6'8"	7'6"	32'0"	22'6"
60,000	5	5'8"	7'6"	43'0"	25'0"
80,000	5	6'8"	8'0"	49'0"	30'0"
100,000	7	6'0"	8'0"	59'6"	30'0"
120,000	7	6'8"	8'0"	68'6"	30'0"
140,000	7	7'4"	8'0"	74'0"	32'0"

Figure 4.8-10. Flow distribution for stationery regenerator. (*Courtesy Regenerative Environmental Equipment Co., Inc., Morris Plains, NJ.*)

HEAT RECOVERY EQUIPMENT 75

4.9 COMBUSTION AIR PRE-HEAT SYSTEMS

The pre-heating of combustion air through heat recovery was described in Section 2.4, along with data for calculating the fuel savings that may be achieved. Heat exchangers used for pre-heating combustion air were described in Section 4.8. In this section, burners and control equipment used for combustion air pre-heat and their incorporation into a complete system are discussed. The elements of a simple combustion air pre-heat system are shown in Figure 4.9-1.

Combustion air is supplied by a blower at ambient temperature plus the blower temperature rise, and at the pressure required by the system. The amount of air flow is controlled by an air damper, operated by a motor. The damper motor operation is, in turn, controlled by a temperature controller at the furnace or oven. The controller may either continuously modulate the air flow, or alternate between high-fire and low-fire conditions. The former is a more accurate method of temperature control; the latter is less complicated and very often provides all the temperature control that is required. Following the damper, a piping connection is provided for the air impulse line. This small pipe conveys the air pressure after the damper to the ratio regulator. The ratio regulator is a simple diaphragm pressure regulator. The combustion air is pre-heated in the heat exchanger and enters the burner air connection.

Gas (or oil) from the supply line passes through a safety shut-off switch, electrically operated by a flame-sensing device on the pilot flame. If there is no

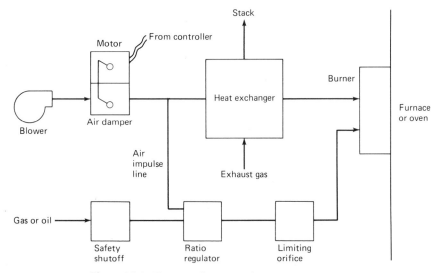

Figure 4.9-1. Elements of a combustion air pre-heat system.

pilot flame, the safety shut-off valve remains closed. The gas passes through the ratio regulator valve, where the gas pressure is regulated to be equal to or to hold a fixed ratio to the air pressure. The limiting orifice is a fixed adjustment of gas flow, set at the initial burner calibration. The ratio regulator, in conjunction with the limiting orifice, maintains a fixed air/fuel ratio at the burner at different air damper settings. The gas is ignited in the burner with a pilot flame. The exhaust gases are passed through a flue into the heat exchanger and the cooled gases are then exhausted out the stack.

Burners for pre-heated combustion air are available with or without attached heat exchangers. Burners with heat exchangers are advantageous where the amount of recoverable heat for an individual burner is adequate, or where the combustion air for each burner is separately controlled. The amount of piping and controls for attached heat exchangers is greatly reduced, and replacement of a defective part does not mean shutting down the entire system. These burners are used on radiant tube atmosphere furnaces where exhaust temperatures are very high, and on forging furnaces and melting furnaces. High-temperature air burners alone are used on furnaces where the pre-heated air comes from a single furnace heat exchanger, installed at the furnace stack or exhaust. Although considerable ducting may be required, the cost of the heat exchanger installation and controls is reduced over the individual heat exchanger system.

A burner with its own heat exchanger is shown in Figure 4.9-2. The flow pattern is shown in Figure 4.9-3.

The heat exchanger is the ceramic cross-flow heat exchanger described in the previous section, mounted on a common base with a nozzle-mix burner. Referring to Figure 4.9-3, combustion air enters the heat exchanger block at the spring-loaded clamping plate, flows across the block, and is pre-heated. It leaves the block and flows directly into the burner. Exhaust gas leaves the furnace through an outlet flue and passes through the block in several passes in the cross-flow direction. The cooled exhaust gas is exhausted by an ejector in the exhaust stack. Although the ejector uses some air flow, the cost of the installation is reduced, since no fan or ducting is required. Ejector air is provided by the combustion air fan. A typical multiple burner installation is shown in Figure 4.9-4.

The elements of Figure 4.9-1 are repeated in Figure 4.9-5, with the exception of the type of gas flow controller. The differential pressure across an orifice plate in the combustion air line is applied across one diaphragm of the ratio regulator. In this system, the ratio regulator is a double-diaphragm pressure regulator. The differential pressure across the limiting orifice appears across the second diaphragm of the ratio regulator used in this system. The balance of diaphragm movement produces a limiting orifice pressure drop in the proper ratio to the combustion air orifice pressure drop. Consequently,

HEAT RECOVERY EQUIPMENT 77

Figure 4.9-2. Burner with integral heat exchanger. (*Courtesy Selas Corporation of America, Dresher, PA.*)

the air and fuel flows are maintained in the proper ratios. The type of control shown in Figure 4.9-1 will cause the air/fuel ratio to become rich at high pre-heat temperatures when properly set at room temperature, wasting fuel. Conversely, it will lean out at low air pre-heat temperatures when set properly at high pre-heat temperatures, wasting fuel again. This is because the control system maintains unheated air and fuel pressures. As the combustion air is heated, it expands and increases in volume; its pressure drop increases across the heat exchanger and the burner. The flow reduces to bring the pressure to the amount being maintained by the blower, causing the rich combustion condition.

Electronic controllers are available to perform the necessary control functions with micro-processors. Measurements are made of all flow rates,

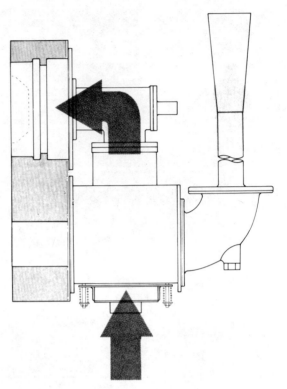

Figure 4.9-3. Flow pattern for burner with integral heat exchanger. (*Courtesy Selas Corporation of America, Dresher, PA.*)

temperatures, and pressures. Signals are continuously sent to the control motor to maintain correct combustion conditions. These systems are useful for multiple burner installations and for programmed furnace operations. The ejector can serve to adjust furnace internal pressure to a neutral value. The ejector air flow is controlled by a damper and drive motor connected to a furnace pressure controller. The ejector flow is adjusted to balance exhaust and supply flows to the furnace. A negative furnace pressure causes infiltration of cold air into the furnace through doors, posts, and cracks. A positive furnace pressure causes exhaust gases to leak through the same openings. Both conditions are wasteful and increase furnace operating cost.

The self-recuperative burner is another type of burner with its own heat exchanger. A drawing of the self-recuperative burner is shown in Figure 4.9-6.

The heat exchanger is built into the burner assembly; combustion air enters the burner assembly in an annular passage around the flue. The air flows around the exhaust plenum and into the burner section counter-flow to the

HEAT RECOVERY EQUIPMENT 79

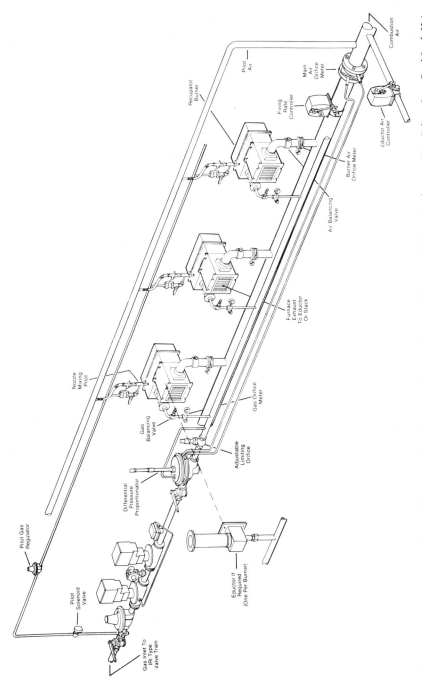

Figure 4.9-4. Multiple installation of burners with self-contained heat exchangers. (*Courtesy Eclipse Combustion Div. Eclipse Inc., Rockford, IL.*)

80 INDUSTRIAL AND COMMERCIAL HEAT RECOVERY SYSTEMS

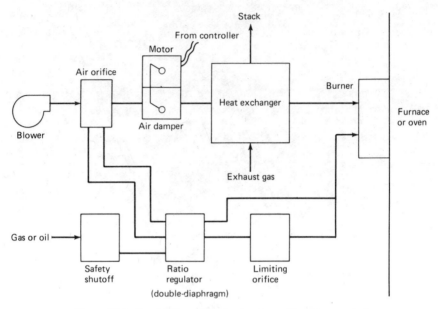

Figure 4.9-5. Combustion air pre-heat system with volume control.

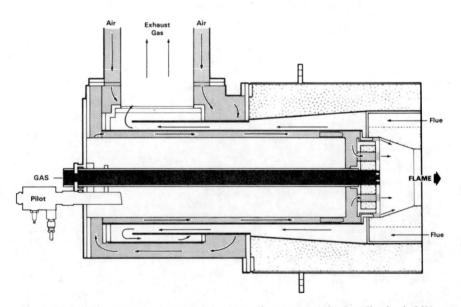

Figure 4.9-6. Self-recuperative burner. (*Courtesy North American Mfg. Co., Cleveland, OH.*)

leaving exhaust gas. The combustion air and gas or oil mix and burn in the burner tile. The exhaust gas reverses direction in the plenum and is drawn out the flue by an ejector. The complete heat exchanger-burner package is mounted in the furnace or oven, eliminating ducts and their associated heat losses and cost. The self-recuperative burner has an effectiveness of 35% to 40% and capacities from 0.5 to 2.5 million Btu/hr. The burner tile diameter varies from 16 in. to 30 in., and the overall burner length varies from 37 in. to 47 in. The control system of Figure 4.9-5 is used with the self-recuperative burner to maintain a fixed air/fuel ratio at various pre-heated air temperatures. Ejector air flow is also controlled to maintain a neutral or slightly positive furnace pressure. The limits of air flow through the burner are defined by two conditions. At the lowest flow, the burner metals will reach a limiting temperature. At the upper flow limit, the burner will reach the limit of its stable operation. The ratio of these flows is the "turndown." The self-recuperative burner has a turndown ratio of 5 to 1. It is also limited in use with sulfur-bearing fuels and drying ovens. In both cases, condensation may occur in the heat exchanger section with resulting corrosion problems.

There are a variety of burners available for use with pre-heated combustion air. One of the more common types is the nozzle-mix burner, shown in Figure 4.9-7. A schematic diagram of the burner is shown in Figure 4.9-8.

The internal burner parts are manufactured from alloy steel and ceramic. The burner will operate with air pre-heated up to 1100° F. The burner can be operated on oil or gas. The fuel and air mix downstream of the air-distribution disc. The length of the flame can be controlled by the size and construction of the disk. The burner is bolted directly to furnace shell, and fires through a port in the refractory furnace lining. The burner, shown in Figure 4.9-8, also includes a refractory tile at the outlet, which serves to improve flame retention by radiation into the combustion area. Burners are equipped with flame detection equipment to operate the electric safety valves if the pilot is not lit, or to shut off the fuel if ignition is not achieved. Oil is atomized either by steam or compressed air. These burners are available in capacities from one million Btu/hr to 30 million Btu/hr. Another type of air pre-heat burner used on radiant tube furnaces is shown in Figure 4.9-9.

Radiant tube furnaces are particularly suited for combustion air pre-heating because of the high exhaust temperatures from the radiant tubes. Since the burners, tube, and exhaust are easily integrated, the type of heat exchanger shown in Figure 4.8-3 has been extensively used on radiant tube furnaces, supplied with its own pre-heat burner. The package includes the ejector, which supplies the draft for the system. No piping or ducting is required for the packaged system, resulting in low installation cost and low operating losses. The application of pre-heated air to a furnace, oven, or process needs to be approached from an overall viewpoint. If the combustion

Figure 4.9-7. *Nozzle-mix burner.* (*Courtesy North American Mfg. Co., Inc., Cleveland, OH.*)

control system is the pressure-regulated system of Figure 4.9-1, it will need to be changed to the flow-regulated system of Figure 4.9-5, or an electronic system. Burner replacement may be necessary, which will require matching of the flame characteristics and turndown of the new burner to the existing burner. Combustion air pressure requirements may require the existing blower to be modified or replaced. Furnace pressure control, if not already

HEAT RECOVERY EQUIPMENT   83

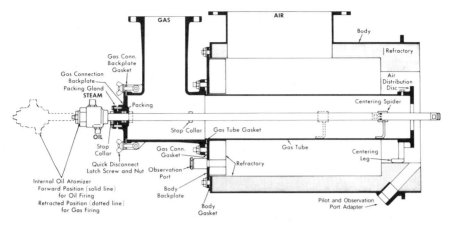

Figure 4.9-8. Schematic diagram of nozzle-mix burner. (*Courtesy North American Mfg. Co., Inc., Cleveland, OH.*)

installed, will be advantageous. Since the burner system will be capable of greater heat release, production throughput increases may be possible at higher furnace or oven temperatures. Finally, operation of the heat exchanger should be checked over the turn-down range of the burner, at start-up, and at shut-down. Care is necessary to avoid condensation and corrosion, or overheating, in the heat exchanger. If either of these conditions can exist, then limits must be placed in the control system.

Figure 4.9-9. Pre-heated combustion air radiant tube burner. (*Courtesy Hauck Manufacturing Co., Lebanon, PA.*)

4.10 GAS/LIQUID HEAT EXCHANGERS

The rate of transfer of heat between a gas and a liquid in a heat exchanger is restricted by the gas heat transfer coefficient, which is much smaller than the liquid heat transfer coefficient. The gas/liquid heat exchanger is constructed with additional heat transfer surface on the gas side to make up for the lower gas heat transfer coefficient. In this way, the overall heat transfer rate is maintained at the design condition. This results in the lowest cost and size heat exchanger to produce a specified heat transfer rate.

The most common form of gas/liquid heat exchanger provides the additional gas side heat transfer surface as fins on the gas side. A finned-tube coil is shown in Figure 4.10-1.

Liquid passes through metal tubes arranged horizontally across the face of the heat exchanger. Gas passes horizontally across the tubes. Its flow direction is perpendicular to the face of the tube row. For a single row of tubes, gas and liquid flow in cross-flow directions. For a multiple-row heat exchanger, gas and liquid flow in counter-flow directions.

The moderate temperature finned-tube coil used in make-up air applications or process applications up to 400° F, has copper tubes 0.625 in. to 1.0 in. outside diameter. The fins are made from thin gauge copper or aluminum. The

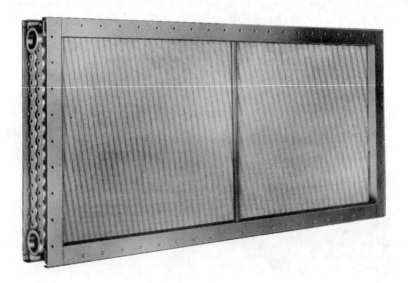

Figure 4.10-1. Finned-tube coil gas/liquid heat exchanger. (*Courtesy The Trane Co., LaCrosse, WI.*)

plate fin coil has rectangular aluminum fins with holes punched in the fins for the tubes. The plate fins can be flat or corrugated. The edge of the hole is extruded to form a collar. The width of the collar establishes the fin spacing. After a row of plate fins has been stacked on a tube, the interior of the tube is expanded by pushing a hardened steel ball through it. The expanded exterior of the tube locks the plate fin collars into place and provides a good metallic seal and contact between the collar and tube. The seal is important to prevent corrosion of the metallic contact. The metallic contact provides good heat conduction from the tube wall to the collar to the plate fin itself. The plate fin transfers heat to the gas stream. The round fin coil has several methods of attaching the fins to the tube. Copper fins are wound onto the tube and dip-soldered into place. The solder assures a good conductive heat path from the tube to the fins. Aluminum round fins are made with collars, and held on by expanding the tube interior. High-temperature coils are made from steel or stainless steel, requiring welding of the fins to the tubes. The ends of the tubes are connected by U-bends brazed into place, or bushed into manifolds made of cast iron or steel tubing. The tube ends are bushed into the cast iron manifold and expanded with clean-out plugs opposite the tube ends. They are brazed into place in the steel tubing manifold.

Baked phenol-formaldehyde coatings are used on finned-tube heat exchangers to increase their resistance to corrosion by certain chemicals. The coating is applied by multiple dipping and baking of the entire coil. The coating is limited to operating temperatures below 450° F. It is not recommended for atmospheres containing strong alkalies and oxidizers, wet bromine, and chlorine and bromine, in concentrations greater than 100 parts per million. The coating must not be scratched, and is weakest at sharp corners and fin edges. The heat transfer properties of the coil are not affected by the coating. The cost of the coated coil will normally be much less than a stainless steel coil, and with better heat transfer properties.

The fin spacing varies according to the expected contaminants in the gas stream. For extremely dirty service where frequent cleaning is expected, fin spacings as low as 5 fins per in. are available. For clean exhaust or supply air, fin spacings as high as 15 fins per in. are available. The fin spacing and the number of rows of coils in the gas flow direction will determine the gas pressure drop across the heat exchanger. The liquid flow path in the heat exchanger can be chosen to satisfy the pressure drop or quantity flow requirements. Normally, the liquid is arranged to pass through all the tubes of a row at once, or a single circuit coil. A double circuit coil divides the liquid flow between two rows of tubes, further reducing the liquid pressure drop, or increasing the flow for the same available pressure drop. For smaller flow rates, partial-circuit coils are made, splitting the flow between upper and lower halves of a coil for a half-circuit, or in third for a one-third circuit. In this

86 INDUSTRIAL AND COMMERCIAL HEAT RECOVERY SYSTEMS

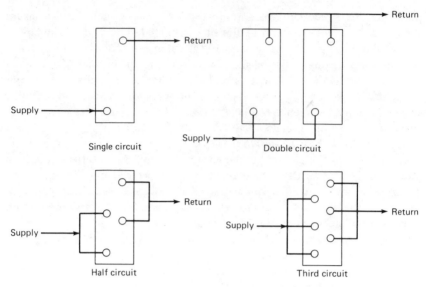

Figure 4.10-2. Finned-tube coil circuit arrangements.

way, smaller flow rates are accommodated at the same pressure drop in fewer rows. These circuits are illustrated in Figure 4.10-2.

The coils are mounted in galvanized steel frames. The frames are supplied with punched flanges for direct mounting to the duct, or to slide into a frame already part of the duct. The latter is most suitable when cleaning may be necessary. Coil widths over 12 ft are available, with face heights over 3 ft. Coils can be stacked to make coil assemblies of as many rows as necessary. Coils can also be stacked to achieve large face heights.

Finned-tube coils on air makeup service must be protected against freezing. If the liquid inside the coil is in a closed system, antifreeze can be added to the liquid. An ethylene glycol-water solution, 30% by volume, will freeze at 0° F. A 50% solution will freeze at −40° F. If the liquid composition cannot be changed, face-and-bypass dampers are fitted to the coil. When a thermostat after the coil face senses a temperature of 35° F, the face damper closes and the bypass damper opens, protecting the coil. Similarly, the coil is protected from exceeding its operating temperature when used in the exhaust stream. Using a face-and-bypass coil with a thermostat set at the high limit of coil operation, the damper operation will prevent coil damage in case of excessive exhaust temperature. Pump failure can cause excessive pressure in the tubes. A pressure switch or thermostat in the piping can also operate the face-and-bypass damper system.

The coil-loop or run-around-loop is a system consisting of two finned-tube

HEAT RECOVERY EQUIPMENT 87

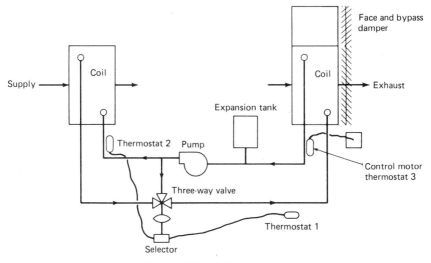

Figure 4.10-3. Coil loop system.

coils. One coil is placed in the exhaust system, the other is placed in the supply system. The coils are connected with a piping loop, circulating pump, and three-way valve, as shown in Figure 4.10-3.

The circulating pump maintains a steady circulation of the glycol-water solution between the supply and exhaust coils, transferring heat from the exhaust to the supply coil. There is no freezing problem in the supply coil, but condensate can freeze in the exhaust coil. If thermostat 2 senses a solution temperature below 35° F, the three-way valve begins to open. Part of the solution bypasses the supply coil and is not cooled, raising the solution temperature. If the supply is heated above the desired temperature by the solution, thermostat 1 will also bypass part of the solution to avoid excessive supply heating. Where a danger exists of boiling the liquid if too much liquid is bypassed, thermostat 3 will operate face-and-bypass dampers at a temperature below the boiling point of the liquid.

The coil-loop is especially advantageous where the source and use for recovered heat are separated by considerable distance, walls, and obstructions, or where little space for the equipment is available. Package units are available for interior or rooftop mounting, which include the coil, fan, and filters. These are particularly useful where the recovered heat is used for heating makeup air. A package unit is shown in Figure 4.10-4.

The effectiveness of the finned-coil increases for the number of rows and the amount of fin area. The effectiveness can reach above 90%, but it is usually in the range of 60%. The gas pressure drop is normally in the range of 0.2 to 1.0 in. water, and the liquid pressure drop is up to 15 ft water.

Figure 4.10-4. Outdoor package air make-up supply unit. (*Courtesy The Trane Co., LaCrosse, WI.*)

The use of a gas/liquid heat exchanger using water or water-glycol mixtures is limited to temperatures compatible with the allowable system pressure. At 400° F, water has a vapor pressure of 232.6 psig, certainly beyond the range of normal finned-tube heat exchangers. Higher-temperature applications can be handled using special heat transfer fluids. These fluids are either based on petroleum or synthetic compounds. Dowtherm* G has a usable range of 20° F to 700° F, with vapor pressure of 30 psig at 700° F. Therminal** 55 has a usable range of −5° F to 600° F with a vapor pressure of 108 psig at 600° F. Some precautions are necessary with these fluids. All joints must be carefully sealed, welded, or brazed, because the liquids will escape from joints leak-tight to water. Because of the cost of these liquids, leaks are not allowed to continue, and system maintenance must be excellent. A liquid should be chosen which will not freeze at the lowest ambient temperature. Because of thermal degradation, liquid replacement is necessary on a continuous basis, and should be taken into account when calculating the cost saving of the heat recovery system.

Another type of gas/liquid heat exchanger is illustrated in Figure 4.10-5.

The system makes use of two spray towers, each filled with a porous packing material. One spray tower handles exhaust air; the other spray tower handles supply air. The two towers are connected by a pumping loop consisting of two circulation pumps, and a solution heater. The circulating

---

*Trademark of Dow Chemical Co.
**Trademark of Monsanto Industrial Chemical Co.

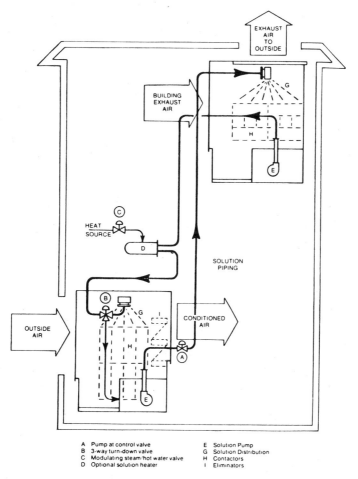

Figure 4.10-5. Direct contact gas/liquid heat exchanger. (*Courtesy Midland Ross Corp., New Brunswick, NJ.*)

liquid is a lithium chloride solution in water, which is hygroscopic. The temperature of the solution is controlled by the solution heater. The solution is also a bacteriocide, and is claimed to remove 95% of bacteria passing through the equipment.

In winter operation, the cold, dry supply air is heated and humidified by the solution, while the moist, warm exhaust air is cooled and dried. If more moisture is given up than received, additional water must be added to the solution. If the supply air temperature is below freezing, or if additional temperature is required at the supply tower outlet, then heat is supplied by the solution heater.

90 INDUSTRIAL AND COMMERCIAL HEAT RECOVERY SYSTEMS

In summer operation, the cold, dry exhaust air is heated and humidified by the solution. The hot, moist supply air is cooled and dried by the solution. No heating or additional water is required. Effectivenesses of 60% to 65% are claimed for this system. Units are available for accomodating to 104,000 cfm.

This direct contact heat exchanger is particularly valuable for use in hospitals and sterile facilities, because it completely separates exhaust and supply flows. The solution will not carry over infectious bacteria, and the heat exchanger will recover both sensible and latent heat. The equipment is relatively large, and does require extensive piping, pumps, and related controls. Another direct contact gas/liquid heat exchanger is shown in Figure 4.10-6.

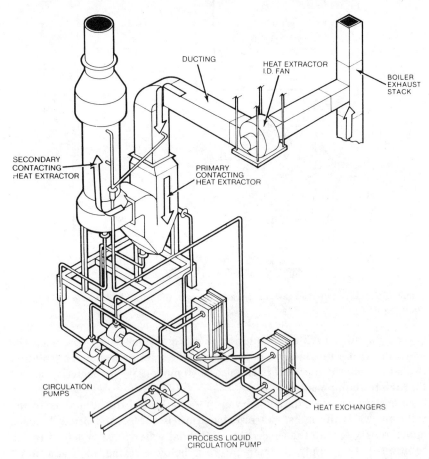

Figure 4.10-6. Direct contact gas/liquid heat recovery heat exchanger. (*Courtesy Heat Extractor Corp., St. Johnsville, NY.*)

A high temperature induced draft fan pulls hot exhaust gases from a boiler, furnace, or process and pushes them into the primary contacting chamber. Water contact in the chamber reduces the gas temperature, producing heated water. The chamber is made of stainless steel to withstand the high entering gas temperatures and corrosive attack from exhaust gas constituents. The heated water is pumped to a plate-type liquid/liquid heat exchanger while the exhaust gases pass to a second contacting chamber. Another water contact in the chamber reduces the exhaust gas temperature to less than 100° F. The cooled gases pass through the water demister and out the stack. The secondary contactor is made from fiberglass to withstand corrosive attack, but at lower exhaust gas temperatures. The heated water is returned to another plate-type liquid/liquid exchanger.

The effectiveness of this heat exchanger can be above 95%, because of the low exhaust gas temperature. Treatment chemicals can be added to the contacting water to reduce the emission of sulfur oxides and control the acidity of the contacting water. About 5% of the contacting water is continuously bled to control solids concentration.

The two plate-type heat exchangers heat water for process, boiler feedwater, wash water for laundries and dye houses, or for any other use. The heated water temperature is about 140° F, depending on the use.

The system is designed with safety interlocks to prevent damage to the contactors if pumps fail or inlet exhaust gas temperatures are excessive. When the induced draft fan shuts off, the exhaust gases pass out the normal exhaust stack. A slide damper is provided to shut off the contactors when work is being done on them.

The primary concern of this exchanger is to provide a carefully engineered system. As in any chemical equipment, pumps, controls, and fans must be properly designed, built, and maintained.

4.11 BOILER ECONOMIZERS

Boilers present an exceptional opportunity for heat recovery. The sources of recoverable heat are the stack gas and the boiler blow-down. The use for recovered heat is the boiler feedwater. Both source and use operate at the same time and in the same cycle. Boiler operation tends to be steady, without rapid changes, and for long periods of time. The boiler stack economizer recovers heat from the stack gas and uses it to heat boiler feedwater. The boiler blow-down economizer uses boiler blow-down to pre-heat incoming makeup water. A schematic diagram of a boiler heat recovery installation is shown in Figure 4.11-1.

A boiler stack economizer is a finned-tube heat exchanger. Its construction differs from the gas/liquid heat exchanger described in the previous section

92 INDUSTRIAL AND COMMERCIAL HEAT RECOVERY SYSTEMS

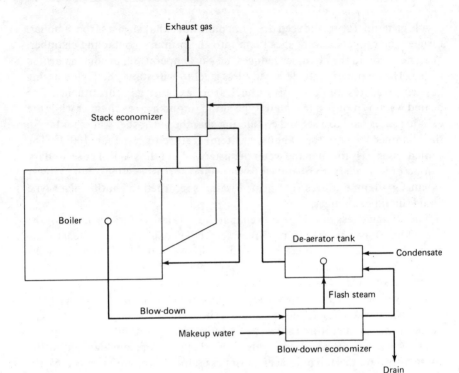

Figure 4.11-1. Schematic diagram of boiler heat recovery installation.

because of the more difficult operating conditions. A typical boiler stack economizer is shown in Figure 4.11-2.

The casing of the economizer is built of heavy gauge steel. Where the economizer is intended for stack mounting, the casing is constructed to support the weight of the stack above it. The economizer intended for use on heavy oil or coal is equipped with a soot blower. The soot blower is a rotating or stationary nozzle. Steam or air is used to periodically blow soot off the heat transfer surfaces. The nozzle assembly may also traverse across the heat exchanger for a more direct cleaning blast. The fins and tubes are commonly made of steel because of the high operating temperatures of stack economizers. The fins are attached to the tubes in a number of ways, depending on the manufacturer. Fins are helically wound and resistance welded or bonded with high-temperature brazing alloys. Fins are also helically wound into grooves in the tube wall, or machined from the tube wall. Fins also can be cast from steel and attached to the tube by welding or brazing. In whatever method is employed, care should be taken to avoid separation of the fin base from the tube wall. Fin spacings up to 5 fins per in. are in use. A cylindrical stack economizer intended for use on smaller boilers is shown in Figure 4.11-3.

HEAT RECOVERY EQUIPMENT   93

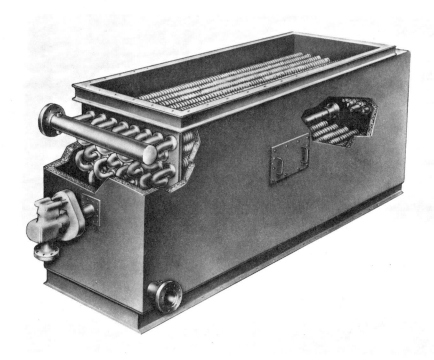

Figure 4.11-2. Boiler stack economizer. (*Courtesy Seton-Scheer, Inc., Medina, OH.*)

The major limitation on stack economizer use is a lower limit on stack gas temperature. Stack gases from sulfur-bearing fuels are not cooled below 350° F as a precaution against the formation of sulfuric acid in the exhaust gas. A very low-sulfur fuel or natural gas may be cooled to 300° F or slightly below, depending on specific conditions. The lower limit of exhaust gas temperature makes the application of stack economizers to low-pressure (15-psig) boilers impractical, since their stack gas temperatures are not normally above 350° F for a well-run boiler. However, higher pressure boilers where stack temperatures can exceed 600° F are well-suited to stack economizers. Boiler size forms another limitation of the use of stack economizers. Small heating boilers are not good applications because of the low flow rates involved, and the restricted yearly use.

Two types of controls are used on boiler stack economizer water flow to avoid acid formation in the economizer, and are shown in Figure 4.11-4a,b. In Figure 4.11-4a, a temperature sensor in the stack controls the amount of feedwater admitted to the stack economizer. When too much feedwater passes through the economizer, the stack gas is cooled below the sulfuric acid dew point. By reducing the feedwater flow, the stack gas temperature is increased

94 INDUSTRIAL AND COMMERCIAL HEAT RECOVERY SYSTEMS

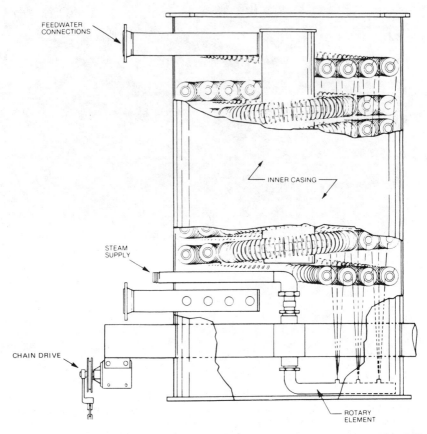

Figure 4.11-3. Cylindrical stack economizer. (*Courtesy Kentube Div., Tranter, Inc., Tulsa, OK.*)

and sulfuric acid condensation is avoided. This method of stack gas temperature control has the disadvantage that at low-fire conditions, or when the boiler is idling, the water flow may be reduced sufficiently to cause steaming in the economizer. Damage may occur to the economizer if the economizer tube pressure suddenly increases and a tube ruptures.

Figure 4.11-4b presents a way to avoid this problem, at the cost of an additional water/steam heat exchanger and steam. The boiler feedwater passes through the water/steam heat exchanger before entering the stack economizer. If the stack gas temperature falls below the pre-set limit, the sensor causes the thermostatic valve to open. Steam passes into the exchanger and pre-heats the feedwater until the sensor is satisfied. The steam condensate is returned to the condensate return system. A second thermostat senses the temperature of the feedwater entering the economizer. If the feedwater

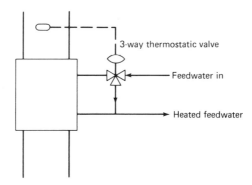

Figure 4.11-4a. Bypass control of water temperature.

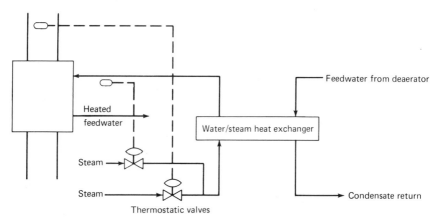

Figure 4.11-4b. Temperature control of water temperature.

temperature is below the limits shown in Figure 4.11-5, sulfuric acid will condense on the tubing.

Below the limiting temperature, the second thermostatic valve opens, preheating the feedwater entering the economizer. Using this method of control, both the stack and the economizer are protected from acid corrosion without the danger of economizer steaming.

Besides stack mounting, boiler economizers can also be mounted on a bypass duct from the stack breeching to the stack. The economizer is bypassed, if repairs are required. Mounting in the breeching or in a manifold from several boilers is also possible. A completely self-contained boiler economizer is shown in Figure 4.11-6. An installation diagram is shown in Figure 4.11-7.

A built-in fan pulls exhaust stack gas through the economizer and out a

96 INDUSTRIAL AND COMMERCIAL HEAT RECOVERY SYSTEMS

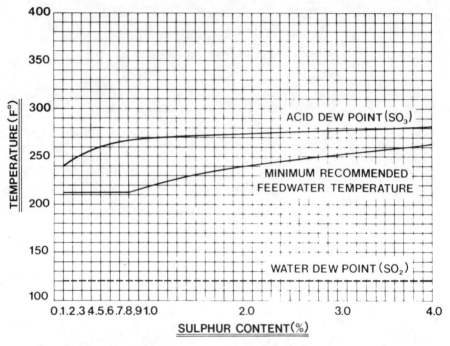

Figure 4.11-5. Relation between sulfur content and feedwater temperature to prevent acid condensation. (*Courtesy Kentube Div., Tranter, Inc., Tulsa, OK.*)

Figure 4.11-6. Self-contained boiler economizer. (*Courtesy Peabody Gordon-Piatt Inc., Winfield, KS.*)

HEAT RECOVERY EQUIPMENT 97

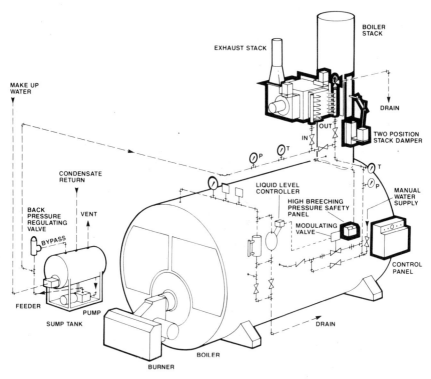

Figure 4.11-7. Installation of self-contained boiler economizer. (*Courtesy Peabody Gordon-Piatt Inc., Winfield, KS.*)

bypass exhaust stack. The fan output is controlled by the boiler stack pressure, maintaining the proper stack draft. Feedwater flow is controlled by the higher liquid level control. The economizer can be by passed by shutting off the induced draft fan. A soot blower is supplied to remove carbon deposits.

In an actual installation, flue gas at 600° F is cooled to 350°, heating feedwater from 200° F to 327° F. The heat recovered is 7,000,000 Btu/hr, saving 11% of the fuel.

Blow-down economizers are used to recover heat from boiler blow-down. Small amounts of feedwater, in the range of 3% to 10% of boiler output, are periodically blown out of the boiler to control the amount of solids in the water. The blow-down contains heat as flash steam from boiler pressure to atmosphere pressure and hot water. To recover this heat, a blow-down economizer similar to that shown in Figure 4.11-8 is used.

The blow-down water at boiler pressure and temperature enters the economizer, and flashes to atmospheric pressure. The flash steam is returned to

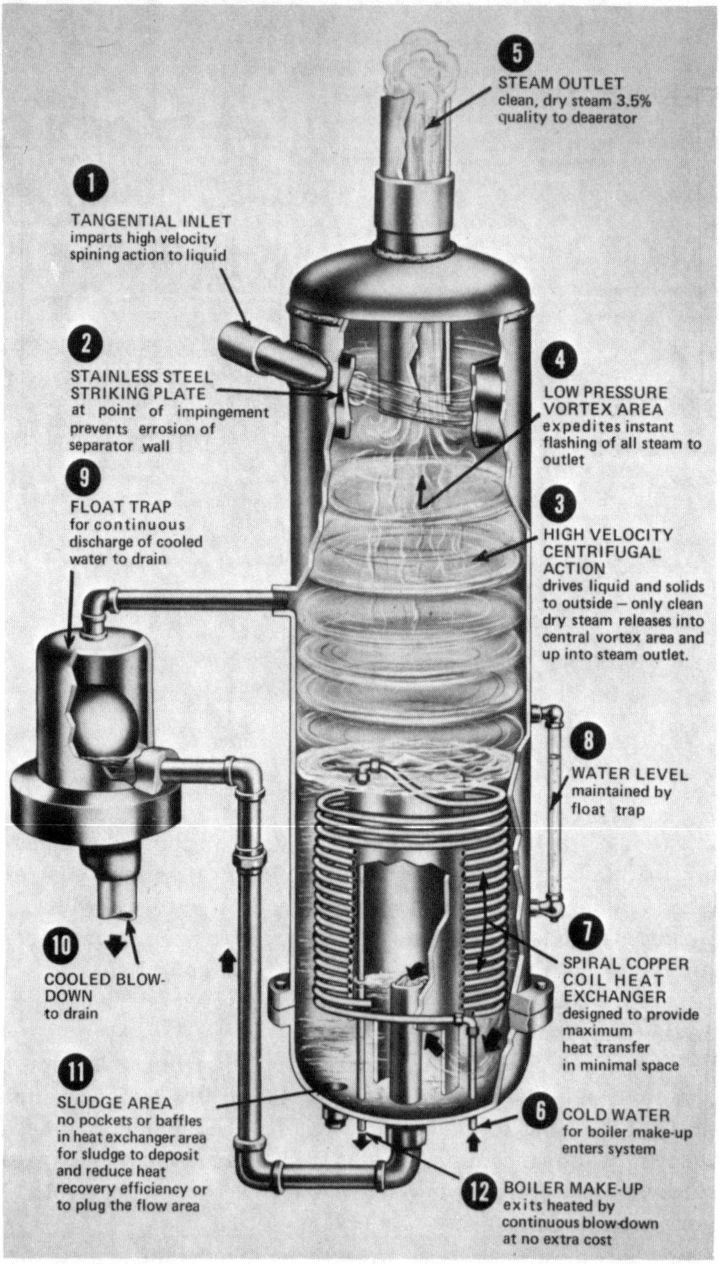

98 INDUSTRIAL AND COMMERCIAL HEAT RECOVERY SYSTEMS

1 TANGENTIAL INLET imparts high velocity spining action to liquid

2 STAINLESS STEEL STRIKING PLATE at point of impingement prevents erosion of separator wall

9 FLOAT TRAP for continuous discharge of cooled water to drain

5 STEAM OUTLET clean, dry steam 3.5% quality to deaerator

4 LOW PRESSURE VORTEX AREA expedites instant flashing of all steam to outlet

3 HIGH VELOCITY CENTRIFUGAL ACTION drives liquid and solids to outside — only clean dry steam releases into central vortex area and up into steam outlet.

8 WATER LEVEL maintained by float trap

10 COOLED BLOW-DOWN to drain

7 SPIRAL COPPER COIL HEAT EXCHANGER designed to provide maximum heat transfer in minimal space

11 SLUDGE AREA no pockets or baffles in heat exchanger area for sludge to deposit and reduce heat recovery efficiency or to plug the flow area

6 COLD WATER for boiler make-up enters system

12 BOILER MAKE-UP exits heated by continuous blow-down at no extra cost

Figure 4.11-8. Boiler blow-down economizer. (*Courtesy Pennsylvania Separator Co., Brookville, PA.*)

the deaerator. The hot water drops to the bottom of the tank where it preheats incoming makeup water in the coiled heat exchanger tubing. The blow-down is discharged through the float trap. Since most areas prohibit the direct discharge of water above 150° F, the unit is designed to cool the water below 150° F, avoiding the necessity of using cold water to mix with the hot water. As an example, a 150 psig boiler blow-down will be cooled to 110° F, saving 80% of the heat in the blow-down.

4.12 WASTE HEAT BOILERS

Waste heat boilers generate steam from hot exhaust gas. The exhaust gas temperature is reduced to a temperature slightly exceeding the boiler operating temperature, dictated by the steam pressure. Low-pressure boilers operate at 15 psig (250° F) and are used primarily for heat. Process boilers operate at 75 psig (320° F), 125 psig (353° F), or above. High pressure boilers for high-temperature processes and power generation operate to pressures of many hundreds of psig. In heat recovery applications, pressures up to several hundred psig are common.

The fire-tube and water-tube boilers are the most common type of heat recovery boilers. A third type, the heat pipe boiler, has recently been introduced. Of the three types, the fire-tube boiler is the most common for package applications and will be described first.

The fire-tube boiler is an outgrowth of the Scotch Marine boiler. A schematic diagram of a simple single-pass boiler is shown in Figure 4.12-1.

The exhaust gas enters the boiler at the right, passes once through the tubes, and leaves the boiler at the left. The exhaust gas transfers its heat across the tube wall to the boiling water surrounding the tubes, producing steam in the space above the boiler water level, or in a separate steam separator drum above the boiler. If a supplemental amount of heat is required, the boiler is built with a tubular combustion chamber at the bottom of the shell, but spaced above the shell bottom to allow space for clean-out of scale accumulation.

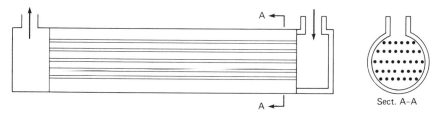

Push thru or pull thru — no supplementary burner

Figure 4.12-1. Schematic diagram of single-pass waste heat boiler. (*Courtesy Eclipse Lookout Co., Chattanooga, TN.*)

100 INDUSTRIAL AND COMMERCIAL HEAT RECOVERY SYSTEMS

Push thru or pull thru — with supplementary burner

Sect. A-A

Figure 4.12-2. Schematic diagram of waste heat boiler with supplemental firing. (*Courtesy Eclipse Lookout Co., Chattanooga, TN.*)

A burner fires into the large tube. The walls absorb the radiant heat and transfer it to the surrounding water. At the end of the first pass, the exhaust gases enter the boiler. They mix with the combustion gases and go through a second pass. Because the mixed gases are at a lower temperature, the second pass tubes are small to enable an increase in velocity and aid in convective heat transfer. The exhaust gases leave at the front of the boiler. If no supplementary firing were used, the exhaust gases would instead be admitted at the front of the boiler, the tube size adjusted depending on the exhaust gas temperature, and the result would be a two-pass boiler. Because the exhaust gas reverses direction in the right-hand plenum, it must be constructed to withstand the direct exposure to hot turning exhaust gas. In the "dry back" construction, the entire chamber is refractory-lined. In the "wet back" construction, the chamber is water-jacketed. The "dry-back" construction allows the use of a rear door in the boiler for access to the tubes, but the refractory must be periodically replaced. The "wet back" construction creates no new maintenance problem, but if a tube replacement is necessary, the repair man may have to crawl up the combustion tube.

A compact multi-pass design is available, where the exhaust gas enters at the center left, as shown in Figure 4.12-3, turns at the right plenum for a second pass, and turns again in a left plenum for a third pass. A divider in the

HEAT RECOVERY EQUIPMENT 101

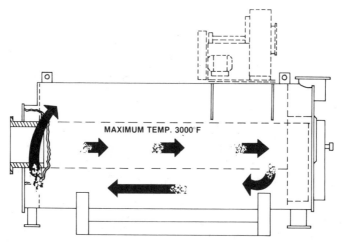

Figure 4.12-3. Schematic diagram of three-pass waste heat boiler. (*Courtesy York-Shipley, Inc., York, PA.*)

right plenum separates the first and third passes. The exhaust gases leave at right side, as shown in Figure 4.12-3.

The choice between the various boiler designs is purely a matter of cost and available space. Since boiler performance depends on heat transfer surface, it is obvious that the desired surface can be provided by various designs. The three-pass design may require more fan horsepower to overcome its additional pressure loss, but this is not a substantial amount.

Fire-tube boilers are constructed like a shell-and-tube heat exchanger. Steel tubes are expanded into the end plates. The shell is rolled, placed around the tubes, and welded into place. The boiler ends, doors, and fittings are added last. Doors allow removal and cleaning of tubes. The boiler is insulated and jacketed. A completed boiler is shown in Figure 4.12-4.

Figure 4.12-4 shows the safety controls required for a fire-tube waste heat boiler. The boiler water level is monitored by a two-position liquid level control. The upper position operates an electric switch connected to the boiler feedwater pump. When the water level falls to the lower operating limits, the feedwater pump is started to supply enough water to bring the level to the upper operating limit. The pump is then turned off. A second lower limit at a lower position is a safety backup for the main liquid level switch. Should the feedwater system fail, the water level could fall low enough to expose the tubes and burn them out. The second or lower limit switch operates fan or damper controls to shut down exhaust gas flow to the boiler if the water level becomes dangerously low. A pressure controller fitted to the boiler is set to maintain the boiler pressure between two points. At the lower pressure point, exhaust

Figure 4.12-4. Fire-tube waste heat boiler. (*Courtesy North American Mfg. Co., Cleveland, OH.*)

gas is permitted into the boiler. At the higher level, exhaust gas is shut off. Modulating controls may also be used. A pressure relief valve on the boiler vents the boiler pressure to atmosphere should the pressure controls fail and allow the boiler pressure to increase to the relief valve setting.

Water-tube boilers are sometimes used in heat recovery applications. Water flows inside the tubes, and the exhaust gases flow outside the tubes, as in the gas/liquid heat exchanger. For high-temperature exhaust gases, the tubes are bare metal, since most of the heat transfer is by radiation. The tubes are finned for convective heat transfer at lower temperatures. Both types of tubes can be used in the same boiler. The essentials of a moderate-sized finned-tube water-tube waste heat boiler are shown in Figure 4.12-5.

Hot exhaust gases pass over the boiler tubes and around circulation baffles. Steam generated in the tubes rises to the steam separator drum. Steam is fed to the piping system from the drum, and water flows down the downcomer to the lower drum, to repeat the process.

Water circulation is by natural convection. Forced convection by pumps can be used to increase the heat transfer rate. The fins are spaced from 3 to 6 fins per in., depending on the amount of particulate present in the exhaust gas. The boiler can be equipped with a soot blower to remove accumulations of

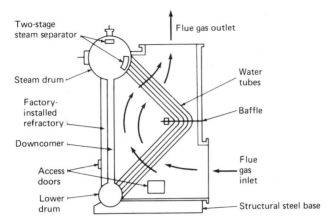

Figure 4-12.5. Essentials of a finned-tube water-tube waste heat boiler. (*Courtesy Deltak Corp., Minneapolis, MN.*)

particulate. A water-tube waste heat boiler is shown in Figure 4.12-6. Water-tube boilers are advantageous for low-pressure drop requirements or where a great deal of contaminants are present. They are not used on smaller applications where a packaged fire-tube boiler is more cost effective.

A heat pipe waste heat boiler is shown in Figure 4.12-7. The heat pipe waste heat boiler is an adaptation of the heat pipe, to produce steam. Hot exhaust gas passes over the finned portion of the heat pipe, vaporizing the internal fluid. The vapor travels up the tube to the bare tube portion inside the steam drum, where it condenses and produces steam. The heat pipes are tilted downward towards the finned end to increase the condensate flow rate, which increases the heat transfer rate of the heat tubes. The design is considerably more compact than standard waste heat boilers, and will produce up to 125 psig steam. The fins are spaced from 2 to 6 fins per in. Exhaust gas from 500°F to 1200°F can be handled at pressure drops of 0.2 in. to 4.0 in. water. The steam drum is provided with double low water level controls, pressure control, relief valve, and high-pressure limit control.

The performance of waste heat boilers is determined by the amount of cooling of the exhaust gas. The boiler pressure and temperature determine the maximum cooling possible. Further limits are set by the necessity of avoiding condensation inside the boiler. For practical purposes, boilers do not operate at less than 350°F outlet exhaust gas temperature. For moderate pressure (125 psig) boilers, the outlet temperature will be on the order of 500°F. Since these temperatures are well above the condensation point of sulfur and chlorine compounds in exhaust gases, chemical corrosion is not a problem in waste heat boilers. If the exhaust temperatures fall too low, then corrosion is

Figure 4.12-6. Water-tube heat boiler. (*Courtesy Deltak Corp., Minneapolis, MN.*)

Figure 4.12-7. Heat pipe waste heat boiler. (*Courtesy Q-dot Corp., Dallas, TX.*)

possible. Waste heat boilers are not normally rated on effectiveness. But as an example, a boiler taking in exhaust gas at 1500° F, feedwater at 250° F, and exhausting the gas at 350° F would have an effectiveness of 92%. Boiler gas pressure drop can vary from 0.2 in. water to 7 in. or 8 in. water, depending on the sizing of the boiler.

The normal method of waste heat boiler control is shown in Figure 4.12-8. This method requires no valve in the hot exhaust gas stream, and automatically returns to the original operating condition if the boiler is taken off stream.

The waste heat boiler inlet is tapped into the exhaust stack of the source of exhaust gas. After passing through the boiler, the gas flow is controlled by a damper and finally exhausted by a fan. The damper and fan are exposed to only cooled exhaust gas and need not be made to withstand extreme temperatures. The position of the exhaust damper is primarily controlled by the boiler pressure control. The control can be a two-position or modulating control. When the boiler pressure is satisfied, the boiler control valve closes and the exhaust gas passes out the original stack. Should the low water cut-off

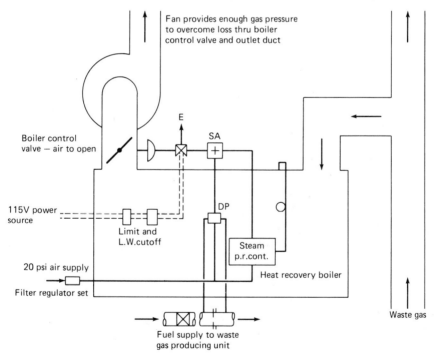

Figure 4.12-8. Two stack with fan method of waste heat boiler control. (*Courtesy Eclipse Lookout Co., Chattanooga, TN.*)

106 INDUSTRIAL AND COMMERCIAL HEAT RECOVERY SYSTEMS

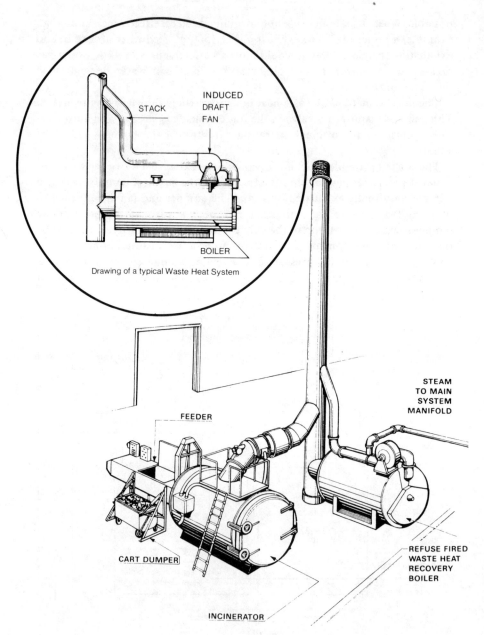

Figure 4.12-9. Incinerator waste heat boiler installation. (*Courtesy York-Shipley, Inc., York, P.A.*)

or high-pressure alarm be actuated, the damper is also closed by the solenoid valve. Finally, if the source of waste heat is a fuel-burning furnace or incinerator, the fuel flow can be translated by a differential pressure cell (DP) to control air pressure.

The control air pressure is sent to a summing amplifier (SA), which selects the proper signal to control the damper. For a simple system, such as an incinerator, the pressure control can be connected to the fan motor control; thus, the damper, fuel supply metering system, and summing amplifier are eliminated. A simple installation of this sort is shown in Figure 4.12-9.

A single stack serves the boiler and incinerator. The boiler intake is well below the boiler stack exhaust to avoid any re-circulation of exhaust gases.

The generation of steam from waste heat is a particularly flexible use for recovered heat. The steam is piped into the existing steam system, avoiding the cost of providing a means for using the recovered heat. The steam can be used for both process and heat. The waste heat boiler provides a back-up for the existing boiler, or a source of emergency steam.

4.13 LIQUID/LIQUID HEAT EXCHANGERS

The liquid/liquid heat exchanger transfers heat between a hot and a cold liquid, across a heat transfer surface. Two types of liquid/liquid heat exchangers are in common use. The shell-and-tube heat exchanger uses a tubular heat transfer surface. The plate heat exchanger uses a flat plate heat exchange surface. Both types are used in heat recovery systems, according to the most feasible choice.

The shell-and-tube liquid/liquid exchanger is similar to the shell-and-tube gas/gas exchanger, but constructed for higher pressures and with more flow circuit options. The exchanger construction is shown in Figure 4.13-1.

The tubes are retained in the tube sheets either by expansion or brazing. The tube sheets are clamped to flanges on the shell and sealed with gaskets. At both ends of the heat exchanger are bonnets, containing fittings for attaching pipes and flow dividers. Pipe fittings are also provided at either end of the shell. Baffles inside the shell distribute the shell flow across the tubes and prevent short-circuiting of the shell flow from inlet to outlet.

For a single-pass heat exchanger, all tube flow passes through the tubes from entrance to exit, while the shell flow passes through the shell once in a direction opposite to the tube flow, or counter-flow. When more heat transfer surface is required, in a given shell length, the tube flow passes through one-half of the tubes, turns in the bonnet, and returns in the other half of the tubes. The bonnet containing the pipe fittings has an internal baffle to divide the entering and leaving flows. Since the two-pass tube flow is both counter-flow and parallel-flow with the shell flow, the heat exchanger performance

108 INDUSTRIAL AND COMMERCIAL HEAT RECOVERY SYSTEMS

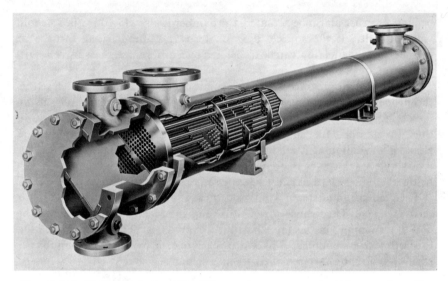

Figure 4.13-1. Shell-and-tube liquid/liquid heat exchanger construction. (*Courtesy Basco Div., American Precision Industries, Inc., Buffalo, NY.*)

reflects the mixed flow pattern. A four-pass heat exchanger is constructed by providing a flow divider in both bonnets. Figure 4.13-2 illustrates these flow patterns.

The fixed tube sheet allows thermal expansion of the tubes only by tube bending, and provides no means of removing the tubes for repair or replacement. The tube sheet at the end of the exchanger remote from the tube flow entrance can be eliminated on two and four pass exchangers. Instead, the tube is bent in the form of a "U", and the tube assembly forms a "bundle." Since the end of the tube floats freely in the shell, it can expand or contract as its temperature changes, and be easily removed from the shell. Another method of construction is to reduce the remote tube sheet diameter so that it fits inside the shell. Sealing methods around the tube sheet are provided to prevent cross-flow between the shell and tube flows. The tube assembly expands and contracts with temperature and may be removed for repair.

These heat exchangers are made from a wide variety of materials, depending on the conditions of use. The standard exchanger will normally have a steel shell, cast iron bonnets, steel tube sheets, and Admiralty metal (copper alloy) tubes. These components are also available in brass, stainless steel, monel, or other alloys. Non-metallic heat exchangers made from graphite or plastics are also available. The choice of materials is made by balancing the cost against expected life and replacement cost.

The plate liquid/liquid heat exchanger resembles a filter press, and is shown

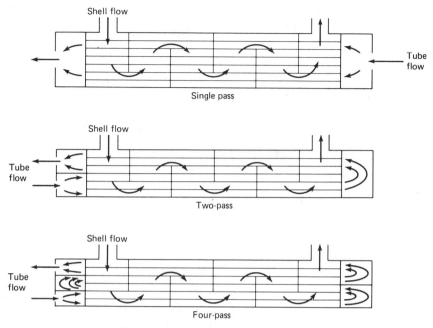

Figure 4.13-2. Shell-and-tube flow patterns.

in Figure 4.13-3. Numerous contoured plates are clamped tightly together in a frame. The edges of the plates and the hole circumferences are gasketed. A typical plate is shown in Figure 4.13-4. The gasketing around the holes is arranged to alternate supply and exhaust flows in the plates, always in a counter-flow direction. The exchanger will have one-pass operation if all the flow is divided equally to all the plates. Multiple-pass operation is possible when the flow is circuited to several paths. The plates are very thin and stamped in a variety of corrugated patterns. The patterns form flow channels on the plate surfaces to produce even flow distribution across the plate and a high velocity, which improves heat transfer. Dimples are sometimes formed in the plates to maintain the proper plate spacing when appreciable pressure differentials exist between the supply and exhaust flows on either side of the plates.

The plate heat exchanger has several advantages over the shell-and-tube heat exchanger. For equal heat transfer surface areas, the plate exchanger is considerably more compact than the shell-and-tube exchanger. Less floor space is required by the exchanger, and no space allowance is required to pull the tubes, as must be provided for the shell-and-tube heat exchanger. If special materials are required because of corrosion, the thin plates use less metal than

110 INDUSTRIAL AND COMMERCIAL HEAT RECOVERY SYSTEMS

Figure 4.13-3. Plate liquid/liquid heat exchanger. (*Courtesy Tranter Div., Tranter, Inc., Wichita Falls, TX.*)

tubes, resulting in a less costly heat exchanger. For normal materials, the shell-and-tube exchanger is less costly than the plate heat exchanger.

The shell-and-tube heat exchanger is advantageous for high-pressure heat exchangers, and for temperatures in excess of 500°F, because of gasket limitations for the plate heat exchanger. Although the plate heat exchanger is easily cleaned, the narrow passages make it more easily to be fouled than the shell-and-tube exchanger. For small exchangers, shell-and-tube heat exchangers are available as stock items, and obtainable from many suppliers.

The performance of liquid/liquid heat exchangers is indicated by the "approach" temperature difference. This is the difference between the entering

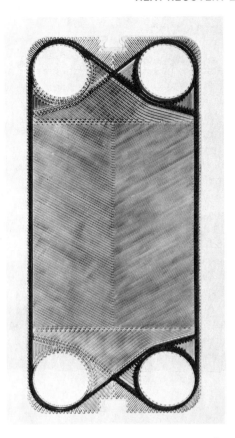

Figure 4.13-4. Heat exchanger plate. (*Courtesy Tranter Div., Tranter Inc., Wichita Falls, TX.*)

exhaust flow and the leaving supply flow. If the approach difference were zero, the effectiveness would be 100%. Many exchangers can achieve an approach of less than 10° F and achieve an effectiveness of over 90%. The greater the heat transfer capability, the smaller the approach temperature and the higher the effectiveness. Economics will determine the most suitable design.

A liquid/liquid heat exchanger should be piped with a bypass and valves to allow maintenance and cleaning. A line filter should be provided if solid materials will be present. A vent valve should be provided to relieve the heat exchanger of air or gases. Adequate floor space should be allowed to remove tubes or plates and carry out required maintenance. A level foundation is necessary to avoid strain on the piping system and the heat exchanger structure. Connections for pressure and temperature gauges are necessary to monitor exchanger performance.

4.14 AIR RE-CIRCULATION

Heat from building air can be recovered without using a heat exchanger. Heated air from beneath the ceiling of a room may be moved back to floor level, or filtered exhaust air may be returned to a building as part of the makeup air system.

In a normal building with a 16-ft ceiling and no unusual manufacturing processes, temperatures under the ceiling can reach 90° F. Where manufacturing operations are present, ceiling temperatures over 120° F are often found. This reservoir of heat is produced by stratified layers of hot air close to the ceiling. Unfortunately, the heat contained in the hot air is lost because the elevated temperature produces additional roof heat loss. The heated ceiling air should be circulated to floor level to heat the working area. For every 1000 cfm of air transferred from a 90° F ceiling area to a 65° F floor level, the heat recovered is 27,125 Btu/hr.

A number of methods are available for re-circulating heated air from the ceiling to the floor. One of these is the ceiling fan shown in Figure 4.14-1.

The fan turns at speeds of 100 rpm to 400 rpm, delivers up to 18,000 cfm, and has a diameter of 48 in. to 60 in. The variable speed motor uses up to 94 watts of power. Slow speeds are used in the winter to bring down ceiling heat and high speeds are used in the summer to increase air circulation in the building. Each fan will cover 1500 ft^2 to 3000 ft^2 of floor space, depending on ceiling height.

Figure. 4.14-1. Ceiling fan. (*Courtesy Robbins and Myers, Inc. Comfort Conditioning Div. Memphis, TN.*)

HEAT RECOVERY EQUIPMENT 113

Figure 4.14-2. Vertical ducted ceiling fan. (*Courtesy Chase Industries, Inc., Cincinnati, OH.*)

Another type of re-circulation system is a vertical duct ceiling fan shown in Figure 4.14-2.

The duct is hung from a ceiling truss. The plastic duct is trimmed to end 3 ft from the floor. The fan at the top pulls 250 cfm to 400 cfm from the ceiling and delivers it directly to floor level. The motor uses 40 watts of power. Each unit covers 2000 ft^2 to 3000 ft^2 of floor space. It is available with a thermostat to turn on the fan when the ceiling temperature exceeds the pre-set value. The duct is 8 in. in diameter.

A third type of re-circulation fan mixes outside makeup air with ceiling air and pushes the mixed air towards the floor. It consists of a wall fan connected to a long horizontal perforated plastic tube. The wall fan pulls in outside air and pushes it out through downward facing holes along the tube length. The

Figure 4.14-3. Ceiling tube fan. (*Courtesy American Coolair Corp., Jacksonville, FL.*)

jets of makeup air mix with ceiling air and set up a re-circulation flow pattern. The tube fan is shown in Figure 4.14-3.

Tube lengths up to 400 ft are available, with fans up to 1 hp. The maximum air delivery is 10,200 cfm.

Another type of re-circulation fan is a floor-mounted fan, shown in Figure 4.14-4.

An internal fan pulls air from the floor and exhausts it at a 9-ft height. A toroidal-shaped flow pattern around the unit traps ceiling air, and brings it to floor level. The largest unit is 26 in. by 52 in. by 9 ft high, has a 3/4-hp motor, and circulates 13,000 cfm. It will serve a floor area of 12,000 ft^2. These units are also available with electric, gas, and steam and hot water heating units for supplemental heat. The manufacturer recommends providing enough capacity to re-circulate the room volume two to four times an hour.

Air can also be re-circulated from many sources of clean exhaust air. If process air is exhausted through a dust collector, it can be brought back into the process or the building. Precautions should be taken not to direct the air directly onto workers. A back-up filter should be provided in case of a failure of the dust collector filter bag. Exhaust air from a paint spray booth can be brought back inside the booth, providing enough outdoor air is added to stay

HEAT RECOVERY EQUIPMENT 115

Figure 4.14-4. Floor-mounted re-circulation fan. (*Courtesy Heat Triever Systems, Fairport Harbor, OH.*)

below explosive limit regulations. Exhaust air from buildings can also be used for process or warehouse ventilation.

4.15 REFRIGERATION HEAT RECOVERY

Heat can be recovered from refrigeration equipment in the form of hot air or hot water. Heat normally lost from the air-cooled condenser can be recovered by a finned-tube coil, or liquid/liquid heat exchanger, operating in conjunction with the air-cooled condenser. A simplified refrigeration system and pressure-enthalpy diagram are shown in Figure 4.15-1.

116 INDUSTRIAL AND COMMERCIAL HEAT RECOVERY SYSTEMS

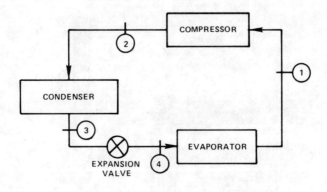

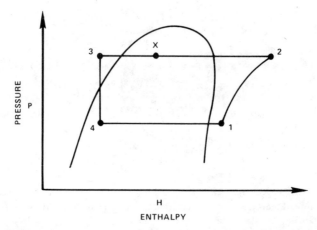

Figure 4.15-1. Refrigeration system schematic and pressure-enthalpy diagram. (*Courtesy The Trane Co., La Crosse, WI.*)

The cooled refrigerant gas at state 1 enters the compressor. It is compressed and heated by the compressor to state 2. The condensor cools the hot refrigerant gas, removing its superheat, and condenses it to a liquid at state 3. The liquid pressure drops through the expansion valve to state 4, and the liquid/gas mixture is evaporated in the evaporator, absorbing heat, and returning to state 1 as a vapor. The maximum amount of heat available for heat recovery is represented by the enthalpy difference between points 2 and 3. For heat recovery purposes, a finned-tube coil or liquid/liquid heat exchanger is connected in series or in parallel with the existing condenser. The heat exchanger is sized to remove typically 75% of the available heat, represented by point x of Figure 4.15-2. The remainder of the heat is removed in the existing air-cooled condenser. The actual position of point x is determined by the amount of heat required: the heat exchanger design, the type of refrigerant, and the operating conditions.

The schematic diagram of the refrigeration heat recovery system is shown in Figure 4.15-2 for a series system. In a parallel system, the heat exchanger discharge follows the air-cooled condenser.

The hot compressed gas flows through the solenoid valve, SLV, to the heat exchanger, where it partially condenses. The remainder of the condensation occurs in the air-cooled condenser. When no heat recovery is called for, the SLV is closed, pressure valve B opens, and the air-cooled condenser removes all the heat. Check valve C prevents back-flow into the heat exchanger when the SLV is closed. Check valve D prevents back-flow into the condenser from the receiver. If the compressor discharge pressure is too high, valve B modulates to the open position to relieve the heat exchanger. If the receiver pressure is too low, pressure valve A opens, admitting more flow to it. The amount of heat recovered is controlled by the pressure difference between the set points of valves A and B.

The heat recovery unit should not be sized to cause excessive condensation or sub-cooling. Excessive condensation will lower the pressure before the expansion valve, resulting in low compressor suction pressure and system capacity reduction. The compressor efficiency will be reduced and slugging of liquid may occur at the compressor inlet; this slugging can also damage the compressor.

Refrigeration heat recovery is used where air conditioners or refrigeration

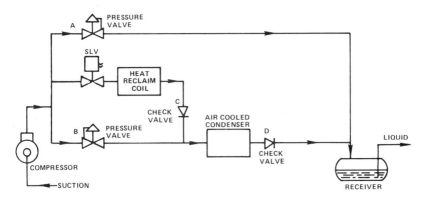

Figure 4.15-2. Refrigeration system schematic diagram with heat recovery. (*Courtesy The Trane Co., La Crosse, WI.*)

machines operate on a fairly continuous basis, and a simultaneous need exists for heat. Supermarkets have air conditioners and freezer cases which operate continuously, and a simultaneous need for building heat and hot water. Hotel restaurants and motels are in the same category, expecially in warm climates. Computer rooms are constantly air conditioned, and building heat can be used during the heating season.

The amount of heat recovered depends on the actual system design. Warm air heating capacity can be estimated from the assumption of a 75% recovery of the 12,000 Btu/ton of rejected refrigeration plus a 15% added compressor heat, or 10,350 Btu/ton of refrigeration/hr. Actual hot water heat recovery units will heat about 10 gallons per hour of water from 65° to 130° F for every ton of refrigeration, amounting to a 36% recovery of available heat.

Finned-tube coils are used for heating air with the recovered heat. These coils are available with multiple circuits when several refrigeration or air conditioning systems are piped to a common heat recovery system. The coil can be installed in a duct next to the heating coil. The heating coil is used if supplemental heat is required. The coil can also be installed in a fan-coil air makeup unit, along with the normal heating coil. The heat recovery coil is equipped with face-and-bypass dampers to protect the coil against frosting, and to reduce flow through the coil if low temperatures may cause excessive condensation.

Two types of hot water heat recovery units are available. The flow-through type is shown in Figure 4.15-3.

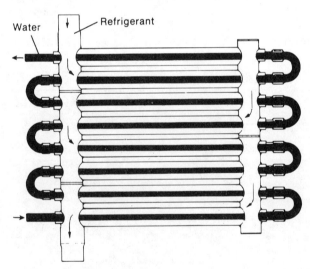

Figure 4.15-3. Flow-through water heat recovery coil. (*Courtesy Enco 2000, Inc., Marietta GA.*)

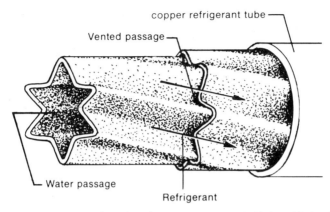

Figure 4.15-4. Vented tube coil construction. (*Courtesy Enco 2000, Inc., Marietta, GA.*)

The refrigerant and water flow through the tubes are in opposite directions, for counter-flow operation. Three tubes form the coil structure, as shown in Figure 4.15-4.

Water flows through the copper central tube. Refrigerants flow between the steel outer tube and the roll-formed helical copper central tube. Heat transfer occurs across the central tube. The helical grooves provide additional turbulence to increase heat transfer, and separate the water and refrigerant. If a leak occurs in the coil, it is vented from the central tube, preventing contamination of the water should a leak occur also in the water tube. A small pump circulates water through the coil. The steel enclosure is insulated with fiberglass. An 11.6-gpm unit is 22 in. high by 42 in. long by 12 in. deep. Hot water can be heated to 150° F. The water is used directly from the coil or stored in a hot water storage tank for use in periods of high consumption. An auxiliary heater in the storage tank will supplement recovered heat, if necessary.

A second type of hot water heat recovery coil is shown in Figure 4.15-5.

The heat transfer surface is a stainless steel double plate, spot-welded at intervals and internally expanded. The refrigerant passes through the expanded passages and is partially condensed, heating water in the tank. The expanded surface is wrapped around the exterior of the storage tank. If additional storage capacity is necessary, auxiliary storage tanks may be used. Storage capacity of 50 to 360 gallons is provided.

4.16 CENTRIFUGAL CHILLERS AND HEAT PUMPS

Heat is recovered from centrifugal water chillers in the same way as the refrigeration heat exchangers of Section 4.15, except that the heat exchanger

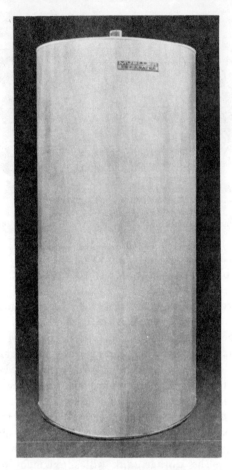

Figure 4.15-5. Hot water storage heat recovery coil. (*Courtesy Paul Mueller Co., Springfield, MO.*)

is built into the water chiller. The basic water chiller cycle is shown in Figure 4.16-1.

In the chilled water circuit, water is chilled from 54° F to 44° F while passing through the chiller. The chilled water coil raises the water temperature back to 54° F, absorbing heat from the air conditioning system. Refrigerant is compressed, condensed, and expanded in the conventional refrigeration cycle. The condenser heat is rejected to a cooling tower. Theoretically, it could also be rejected to a heating coil in the cooling water circuit to recover its heat. This is not actually carried out because the contamination from the cooling tower would greatly increase maintenance in the heating coil. Instead, the heat

HEAT RECOVERY EQUIPMENT   121

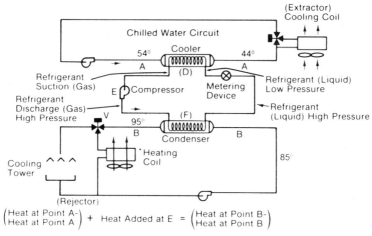

Figure 4.16-1. Basic water chiller cycle. (*Courtesy McQuay-Perfex, Inc., Staunton, VA.*)

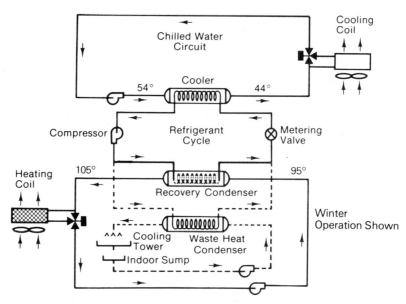

Figure 4.16-2. Schematic diagram of dual condenser heat recovery cycle when heating. (*Courtesy McQuay-Perfex, Inc., Staunton, VA.*)

122 INDUSTRIAL AND COMMERCIAL HEAT RECOVERY SYSTEMS

recovery system is separated from the cooling tower system by the use of two condensers, or two tube bundles in a single condenser. The principle of operation is shown in Figure 4.16-2, and an actual unit is shown in Figure 4.16-3.

The circuit of Figure 4.16-2 includes two condensers, the recovery condenser and the waste heat condenser, instead of the single condenser shown in Figure 4.16-1. The recovery condenser is connected in parallel with the waste heat condenser in the refrigeration cycle, but gives up its heat to a use for recovered heat instead of the cooling tower. A thermostat senses the refrigerant temperature leaving the recovery condenser. If it is excessive, the waste heat condenser and water tower circuit is brought into use. This condition for summer use is shown in Figure 4.16-4.

The recovery condenser discharges at 105°F, while the waste heat condenser discharges at 95°F. The higher water temperature increases the amount of heat available from the heat recovery circuit, but produces a lower compressor efficiency. The penalty of lower compressor efficiency is less during non-heating seasons for heating applications. Process applications for waste heat will benefit from heat recovery throughout their operating periods.

The use of recovered heat from heat recovery water chillers can be greatly increased by the provision of storage capacity for chilled or heated water. In an office building, often both heating and cooling will be required at the same

Figure 4.16-3. Dual condenser centrifugal water chiller. (*Courtesy McQuay-Perfex, Inc., Staunton, VA.*)

HEAT RECOVERY EQUIPMENT

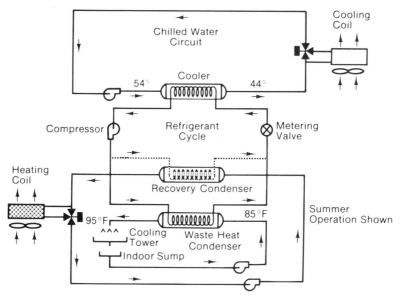

Figure 4.16-4. Schematic diagram of dual condenser heat recovery cycle when not heating. (*Courtesy McQuay-Perfex, Inc., Staunton, VA.*)

time. On weekends, only heating will be required, since the offices will be unoccupied. Excess hot water produced by the recovery condenser during occupied periods can be stored, and used to supply heat during unoccupied periods. The heated water can also be used to false-load the compressor, by blending hot and chilled water, to produce additional heat. Chilled water may also be stored during off-peak hours at night or on weekends. The chilled water may be used during peak hours for building cooling, and the hot water used for building heat. A heat pump can also be used to produce heated water. The heat pump takes water from a source and cools it in the evaporator. Process or building water is heated in the condenser at the same time. The heat pump is non-reversible, used always to produce hot water. The cycle is shown in Figure 4.16-5.

The refrigeration cycle contains the usual condenser, pressure reducing valve, evaporator, and compressor. A flash collector tank bleeds off a portion of the refrigerant at an intermediate pressure and returns it to the inlet of the second compressor stage, reducing the amount of refrigerant handled by the first stage and increasing the compressor efficiency. Source water is taken into the evaporator at 95° F, cooled to 85° F, and drained or used in a process. Process or heating water is taken into the condenser at 140° F and delivered at 160° F. These heat pumps have a coefficient of performance (COP) ranging

How It Works

Using the non-reversible heat pump principle, the Templifier recovers low grade waste heat in the temperature range of 50°F to 140°F (10°C to 60°C) and amplifies it to higher, usable temperature levels. The waste heat from the source water is absorbed in the heat pump evaporator by the unit's working fluid which is then increased in temperature and pressure by the compressor. From here it goes to the condenser where this heat is transferred to a delivery fluid for providing useful heat at temperatures up to 220°F.

Where To Use

- Typical uses: Heat process fluids, process make-up water and hot service water, plus space heating, boiler and other feed water heating.

Typical waste heat sources:

In-Building Sources

- Air Conditioning Equipment Cooling Water
- Overhead Vapors from Distillation Processes
- Warm Water Effluent from Plant Processes
- Refrigeration Equipment Cooling Water
- Air Compressor Cooling Water
- Electric Welder Cooling Water
- Extruder Cooling Water
- Injection Molder Cooling Water
- Cooling Tower/Pond Water

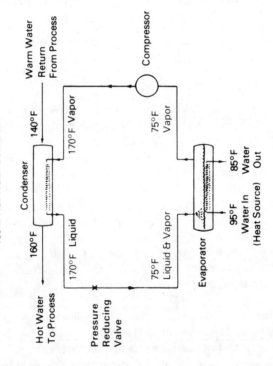

TEMPLIFIER SCHEMATIC DIAGRAM
95°F Heat Source Temperature
160°F Hot Water Delivered

Description

Each Templifier, completely factory assembled, is a packaged industrial heat pump designed to deliver heated water (or fluid as specified) utilizing a waste heat water (fluid) source and electrically powered. Each package includes:

- Centrifugal compressor and motor drive, complete with lubrication system, and where required, interstage equipment,
- Shell and tube source fluid cooler (evaporator),
- Shell and tube delivery fluid heater (condenser), insulated,
- Working fluid (refrigerant) charge,
- Electronic control center containing safety and automatic operating controls,
- Assembly with all interconnecting refrigerant and lube system piping,

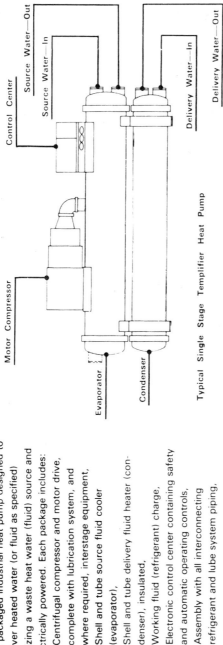

Figure 4.16-5. Heat pump schematic diagram. (*Courtesy McQuay-Perfex, Inc., Staunton, VA.*)

126 INDUSTRIAL AND COMMERCIAL HEAT RECOVERY SYSTEMS

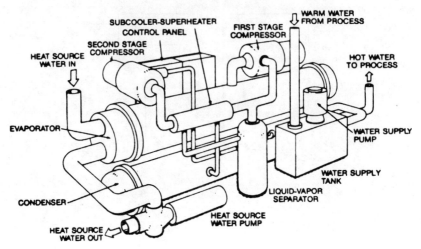

Figure 4.16-6. Heat pump diagram. (*Courtesy McQuay-Perfex, Inc., Staunton, VA.*)

from 3.2 to 5.7, depending on source and output water temperatures. The COP is the ratio of the thermal energy increase of the heated water to the thermal equivalent of the electrical energy consumption of the compressor. The COP signifies that these heat pumps deliver 3.2 to 5.7 times more thermal energy than is put into them. The excess, of course, comes from the source water, and the heat pump actually "pumps" this heat to the higher temperature of the process or heating water. A drawing of the heat pump is shown in Figure 4.16-6.

These heat pumps are supplied in a number of sizes and capacities, as shown in Figure 4.16-7.

The COP decreases as the temperature difference between the source and hot water increases. Source water from 60° F to 120° F can be used. Hot water temperature is limited to 220° F. Source water can be used from processes requiring cooling, such as extruders, welding machines, air compressors, air conditioners, induction furnaces, and injection molders. The need for a cooling tower is eliminated or reduced with these systems. The output water can be used with solar collectors, washing systems, sanitary water supplies, building heat, and many processes. Distillation processes can benefit from this equipment, because the temperature difference between the distillate and the condensate is small enough to be provided by the heat pump.

4.17 ABSORPTION REFRIGERATION FROM RECOVERED HEAT

Absorption refrigeration machines are widely used for producing chilled water from steam or hot water. Absorption machines are available in sizes

PERFORMANCE AND PHYSICAL DATA

Templifier Heat Pump Model	Nominal Peak Heating Capacity and Power Input* Leaving Hot Water/Source Water Temperatures											Dimensions			Approx. Operating Wt.
	120 F/75 F		140 F/80 F		150 F/85 F		170 F/95 F		190 F/100 F		Height	Length	Width	
	Mbh	Kw	Mbh	Kw	Mbh	Kw	Mbh	Kw	Mbh	Kw	in.	in.	in.	Lbs.
TPE 050	2 600	135	2 100	135	1 200	80	1 200	90	1 200	110	72	166	30	7 600
TPF 050**	4 000	205	3 300	210	2 100	140	2 100	160	2 100	195	72	166	30	9 000
TPE 063	4 100	210	3 400	215	2 200	145	2 200	165	2 200	200	78	164	35	11 600
TPE 079	6 400	330	5 200	335	4 000	260	4 000	300	3 700	340	83	166	40	14 700
TPE 100	9 400	485	7 600	485	6 500	425	6 300	475	5 700	525	98	168	48	22 300
TPE 126	22 200	1 140	17 800	1 140	10 000	650	10 000	750	10 000	920	106	200	89	40 000
COP	5.7		4.7		4.5		3.9		3.2				

* Single stage Templifier heat pump heating delivery water to the leaving temperature shown when supplied with sufficient gpm of source water which is cooled to the leaving temperature shown.

** Single stage dual compressor model.

Figure 4.16-7. Heat pump sizes and capacities. (*Courtesy McQuay-Perfex, Inc., Staunton, VA.*)

128 INDUSTRIAL AND COMMERCIAL HEAT RECOVERY SYSTEMS

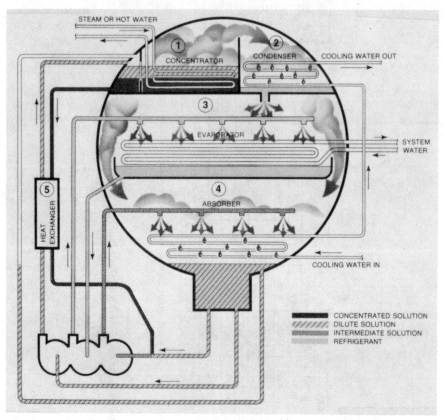

Figure 4.17-1. Basic absorption refrigeration cycle. (*Courtesy The Trane Co., La Crosse, WI.*)

from 3 tons to over 1000 tons of refrigeration capacity, and are used in buildings, food processors, apartment houses, and where a boiler and cooling load are both present. Absorption machines are often used with waste heat boilers to use their steam in the summer. An absorption machine is also available, which uses hot exhaust gases directly for producing chilled water.

Although an absorption machine is quite complicated, the basic cycle can be understood from Figure 4.17-1.

The refrigerant is distilled water and the absorbent is lithium bromide, which has a high affinity for water. At the upper right, the refrigerant vapor is condensed by cooling water from a cooling tower. Because the entire system is under a high vacuum, boiling temperatures are reduced. The pressure in the condenser is about 50 millimeters of mercury, and cooling water temperatures are 90° F to 100° F. The condensed refrigerant passes through an expansion valve where the pressure is lowered to 6 millimeters of mercury. The refrigerant flashes to a vapor and is cooled in the evaporator. Water is chilled in the evaporator for use in the chilled water circuit. Its temperature is reduced from 54° F to 44° F. The warmed refrigerant vapor passes into the absorber, and is picked up by the lithium bromide solution. The heat of absorption is removed by the first pass of the cooling water in the range of 80° F to 90° F. The diluted lithium bromide solution is pumped back to high pressure in the concentrator. Heat is added to the concentrated solution to boil off the refrigerant and repeat the cycle in the condenser. The hot concentrated lithium bromide solution passes through a heat exchanger back to the absorber. The heat is used to pre-heat the diluted lithium bromide solution on its way to the concentrator.

In a heat recovery application, steam from a waste heat boiler or hot exhaust gases are used in the concentrator. The exhaust gases are in the range of 550° F to 1500° F, and are exhausted at 400° F. They may be from combustion products, furnaces, ovens, etc. As an added feature, the condensed hot refrigerant can be bypassed and used directly in a heating

Figure 4.17-2. Heat recovery absorption chiller. (*Courtesy Gas Energy, Inc., Brooklyn, NY.*)

130 INDUSTRIAL AND COMMERCIAL HEAT RECOVERY SYSTEMS

application, since it is at 190° F to 195° F. A heat recovery absorption chiller is shown in Figure 4.17-2. Capacity charts are shown in Figure 4.17-3 and 4.17-4, and the balance of heating and cooling capacities simultaneously available is shown in Figure 4.17-5.

The pressure drop through the heat exchanger is about 6 in. water. These chillers are two-stage absorption machines, in that the diluted lithium bromide solution is heated twice before entering the concentrator. The first heating occurs in the heat exchanger. The second heating occurs in a second-stage concentrator where the hot vapor from the first-stage concentrator is cooled by entering concentrated lithium bromide. Both concentrator stages feed into the condenser.

These units are available in refrigeration capacities from 100 to 1500 tons. As an example, a 500-ton unit will handle 1201 gpm of chilled water or hot water heated from 131° F to 140° F. The cooling water requirement is 2290 gpm at 85° F. The operating weight is 65,000 lb and the size is 272 in. long, 125 in. wide, and 110 in. high.

A small single stage absorption chiller of 25-ton capacity is available for use with steam, hot water, or high temperature exhaust gas, and is shown in Figure 4.17-6.

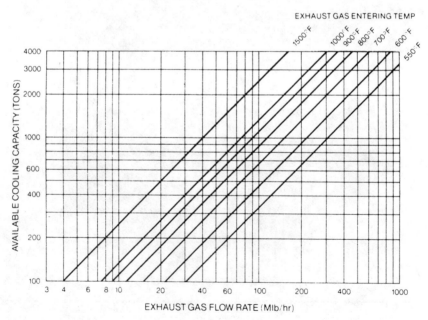

Figure 4.17-3. Heat recovery chiller cooling capacity. (*Courtesy Gas Energy, Inc., Brooklyn, NY.*)

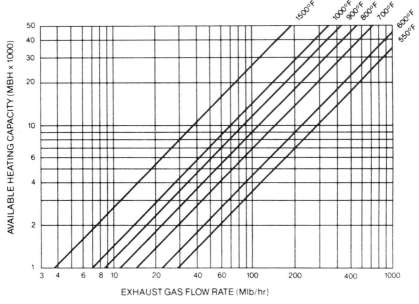

Figure 4.17-4. Heat recovery chiller heating capacity. (*Courtesy Gas Energy, Inc., Brooklyn, NY.*)

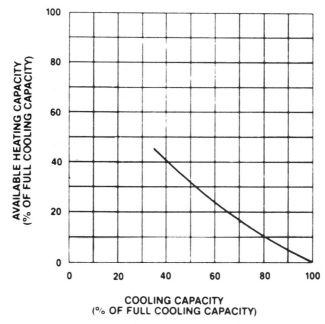

Figure 4.17-5. Balance of simultaneous heating and cooling capacities. (*Courtesy Gas Energy, Inc., Brooklyn, NY.*)

132 INDUSTRIAL AND COMMERCIAL HEAT RECOVERY SYSTEMS

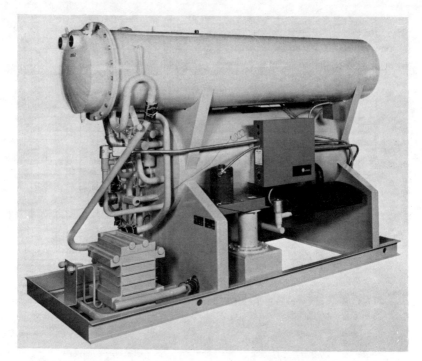

Figure 4.17-6. Small absorption water chiller. (*Courtesy Arkla Industries, Inc., Evansville, IN.*)

Hot water at 170° F to 205° F, or at 225° F, or hot gases in the range of 800° F to 900° F can be used. It produces 60 gpm of chilled water at 45° F. Exhaust gas input is 600,000 Btu/hr, and hot water input is 435,000 Btu/hr. Steam input is the same. It uses 50 gpm to 110 gpm of cooling water at 75° F to 90° F. Operating weight is 3500 lb to 5000 lb, depending on the source of heat. Hot water can be supplied from gas/liquid heat exchangers, laundries, and solar panels. Steam is supplied by waste heat boilers at 15 psig and exhaust gases are used from furnaces and kilns.

4.18 ENGINE AND GAS TURBINE HEAT RECOVERY

Several methods are available to recover heat contained in the exhaust and cooling water from internal combustion engines and gas turbines. These engines are used for power generation, pumping liquids and gases, and as prime movers. They operate for long periods at a steady rate, making them excellent sources for recoverable heat. If oil is used as a fuel, soot-blowing equipment may be necessary to remove carbon accumulation, but otherwise

HEAT RECOVERY EQUIPMENT 133

OPERATION

Exhaust gas enters the heat recovery silencer through connection (1), makes a reversal at the lower end of the unit and passes up over the finned tubes through which the heat is transferred to the water. The gas exits through the exhaust outlet connection (2).

The water enters through connection (3) and from there it enters the lower manifold and passes into the tubes. After absorbing the available heat the water (or steam and water mixture) exits through connection (4).

There are many available combinations of exhaust gas and water flow but the same basic relationships will hold for all.

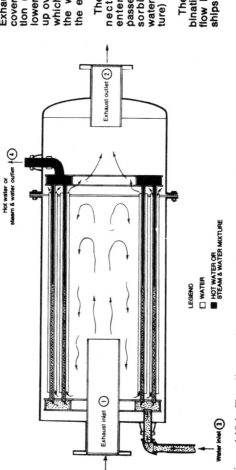

Figure 4.18-1. Flow diagram of heat recovery silencer for reciprocating engines. (*Courtesy Riley-Beaird Div., U.S. Riley Corp., Shreveport, LA.*)

no special conditions exist for recovering exhausted heat, other than avoiding condensation in the heat exchanger.

One of the simplest methods of engine exhaust heat recovery is the heat recovery muffler. A flow diagram is shown in Figure 4.18-1.

Engine exhaust enters the silencer chamber, reverses direction, passes over longitudinally finned tubes, and is exhausted. The water enters through the bottom and passes upwards inside the finned tubes; it leaves at the top through a collection manifold. A view of the heat recovery muffler is shown in Figure 4.18-2.

If hot water is being produced, it is taken directly from the muffler. Steam will be produced as wet steam, which is separated in a separate drum, or in a special separator chamber at the top of the muffler. Other types of mufflers are available containing water in a jacket surrounding the silencing chamber, eliminating the need for tubes and tube sheets, but these require more jacket area.

These exchangers have effectivenesses over 70%, and are capable of reducing exhaust gas temperatures from 1000° F to 400° F while producing steam at 15 psig. An exchanger handling 9000 lb/hr of exhaust gas will produce 1,344,000 Btu/hr at a pressure drop of 5.5 in. water. It will weigh 3200 lb, and have a diameter of 40 in. and a height of 118½ in.

Gas turbine exhaust heat recovery utilizes the same principles, but provides considerably more flow area because of the larger turbine exhaust flow rates. The flow schematic diagram is shown in Figure 4.18-3.

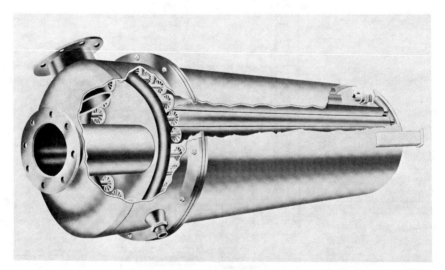

Figure 4.18-2. Cut-away of heat recovery silencer for reciprocating engines. (*Courtesy Riley-Beaird Div., U.S. Riley Corp., Shreveport, LA.*)

HEAT RECOVERY EQUIPMENT 135

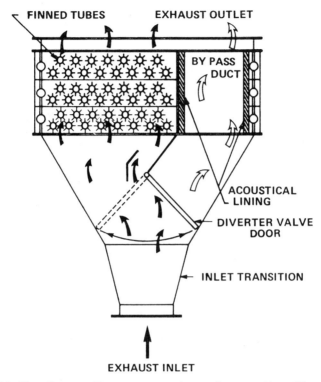

Figure 4.18-3. Flow diagram of heat recovery equipment for gas turbines. (*Courtesy Riley-Beaird Div., U.S. Riley Corp., Shreveport, LA.*)

Exhaust gas enters through the flow inlet, and flows either through the finned tubes, or the bypass duct, depending on the diverter valve position. The diverter valve position is controlled by the outlet steam pressure or liquid temperature. Flow arrangements are available for producing liquid or steam. For the former, all the liquid is single-circuited, passing through all the tubes in series. For steam, the feedwater comes into a lower header and passes up through the tubes in parallel. The water-steam mixture is collected at the top from a header and taken to a water-steam separator drum. The separator drum will produce a minimum 98% quality steam.

A unit handling 65,000 lb/hr of turbine exhaust gas at 1000° F will transfer 12,000,000 Btu/hr to steam at 15 psig. The gas will be cooled to 262° F for an effectiveness of over 90%. This heat exchanger will be $109\frac{1}{2}$ in. long, 93 in. wide and 101 in. high and will weigh 12,770 lb when operating.

These gas turbine heat exchangers are now being used in a heat recovery system to produce electric or mechanical power from engine, gas turbine, or

136 INDUSTRIAL AND COMMERCIAL HEAT RECOVERY SYSTEMS

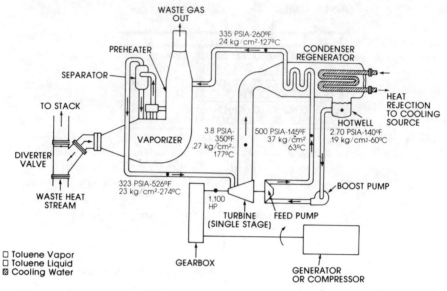

Figure 4.18-4. Turbine heat recovery system. (*Courtesy Sundstrand Corp., Rockford, IL.*)

Figure 4.18-5. Packaged power module for turbine heat. (*Courtesy Sundstrand Corp., Rockford, IL.*)

process exhaust gases. A schematic diagram of the system operation is shown in Figure 4.18-4.

Exhaust gas in the range of 600° F to 1100° F enters the heat exchanger-boiler, and passes through several rows of finned tubes. The fluid inside the tubes, toluene, boils in the rear part of the exchanger. Its vapor is superheated in the front part of the heat exchanger. Toluene is used in the system since its pressure-temperature characteristics fit the requirements for efficient turbine performance at specified exhaust gas temperatures, low toxicity, availability, and cost. The toluene liquid/vapor mixture is separated in a separate drum. The toluene vapor leaves the superheater at 550° F and 283 psig. It expands through an impulse turbine to −11.5 psig and 379° F, and is condensed by cooling water. The liquid toluene is pumped by a boost pump and feed pump back to a pre-heat coil in the condenser, where it is heated from 143° F to 290° F. The feed pump is on the same shaft as the turbine. The liquid toluene goes from the pre-heat coil in the condenser to another pre-heat coil above the vaporizer, and then repeats the cycle by returning to the vaporizer. The packaged turbine, pumps, condenser, piping, and controls are shown in Figure 4.18-5.

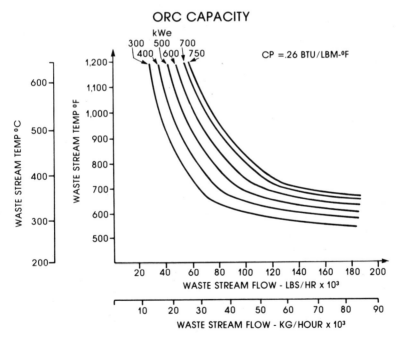

Figure 4.18-6. Turbine heat recovery system performance. (*Courtesy Sundstrand Corp., Rockford, IL.*)

A three-way diverter valve is fitted before the exhaust gas inlet to the vaporizer. The valve position is controlled by the turbine loading to maintain 600 kw or the equivalent 900 hp. Excess exhaust gas is vented to the stack. In case of emergency conditions, all exhaust gas is vented to the stack. The system will operate automatically, maintaining proper conditions, and shut itself down in case of turbine overspeed, improper temperatures or pressures, etc. An idea of its capabilities can be gathered from Figure 4.18-6, relating exhaust gas temperature and flow rate to turbine output.

These units are in use on the exhausts of large reciprocating engines used for power generation, where the turbine power output is added to the grid. They are also in use on a ceramic kiln, a glass furnace, and a solar-powered irrigation project.

REFERENCES

1. Chapman, A. J., *Heat Transfer.* MacMillan, New York, NY. 1960.

5
SIZING AND PERFORMANCE OF HEAT RECOVERY EQUIPMENT

Sizing and performance data are not available for all the equipment covered in Chapter 4 in a form usable for calculation. Many manufacturers perform sizing calculations using computer programs, or do not publish design data. Where available, manufacturers' data are reproduced in this chapter with illustrative examples. Their use will facilitate the estimation of the heat recovered necessary for calculation of heat recovery system economics. Heat recovery equipment sizing information is also presented. Manufacturers must be contacted if the required data are not included in this chapter.

For ease of reference, the equations used in this chapter are repeated below. In the example, numbers after the equation refer to the numbers of the listed equations. Subscripts, nomenclature, and units will be found in their respective listings at the beginning of the book. In these examples, the standard air density is 0.076 lb/ft^3 to agree with manufacturers' data.

EQUATIONS

$$H = C_P W \, \Delta T \qquad (2\text{-}2)$$

$$H = 1.08 \, V_{std} \, \Delta T \qquad (2\text{-}4)$$

$$V_{std} = V \left(\frac{530}{460 + T} \right) \left(\frac{P + P_{std}}{P_{std}} \right) \qquad (2\text{-}5)$$

$$H = 500 \, V \, \Delta T \qquad (2\text{-}6)$$

140 INDUSTRIAL AND COMMERCIAL HEAT RECOVERY SYSTEMS

$$H = \frac{Fu}{100} \tag{2-7}$$

$$V = vA \tag{3-2}$$

$$W = dV \tag{3-3}$$

$$H = W(h_l - h_e) \tag{3-10}$$

$$H = UA \, \Delta T \tag{4-5}$$

$$H = UA \, \Delta T_M \tag{4-6}$$

$$\Delta T_M = \frac{\Delta T_1 - \Delta T_2}{2.3 \log (\Delta T_1 / \Delta T_2)} \tag{4-7}$$

$$E = \left(\frac{\Delta T}{T_{ee} - T_{se}}\right)(100) \text{ or } \left(\frac{\Delta T}{T_{se} - T_{ee}}\right)(100) \tag{4-8}$$

5.1 GAS/GAS CURVED PLATE COUNTER-FLOW HEAT EXCHANGER

Example 1: Unequal flows of dry air.

$$V_e = 14{,}000 \text{ cfm} \quad T_{ee} = 210° F$$
$$V_s = 9500 \text{ cfm} \quad T_{se} = 30° F$$

Desired $E = 70\%$.

1. $V_{\text{std}, e} = \dfrac{(14{,}000)(530)}{(670)} = 11{,}075$ standard cfm. (Equation 2-5)

   $V_{\text{std}, s} = \dfrac{(9500)(530)}{(490)} = 10{,}275$ standard cfm. (Equation 2-5)

2. Based on largest flow, select two units, 48 in. by 48 in. by 6 ft long.
3. Enter Figure 5.1-1 on 48 × 48 line at half $V_{\text{std}, e}$ (5537.5 standard cfm), find $E_s = 70\%$. Find pressure drop on P.D. curve at 0.58 in. water.
4. Enter Figure 5.1-2 for $11{,}075/10{,}275 = 1.08$, find 0.5% efficiency gain and add only to smallest flow: $E_s = 70.5\%$.
5. $\Delta T_s = (0.705)(210 - 30) = 127° F$ (Equation 4-8)
6. $T_{sl} = 30 + 127 = 157° F$.
7. $H = (1.08)(10{,}275)(127) = 1{,}409{,}319$ Btu/hr. (Equation 2-4)

SIZING AND PERFORMANCE OF HEAT RECOVERY EQUIPMENT 141

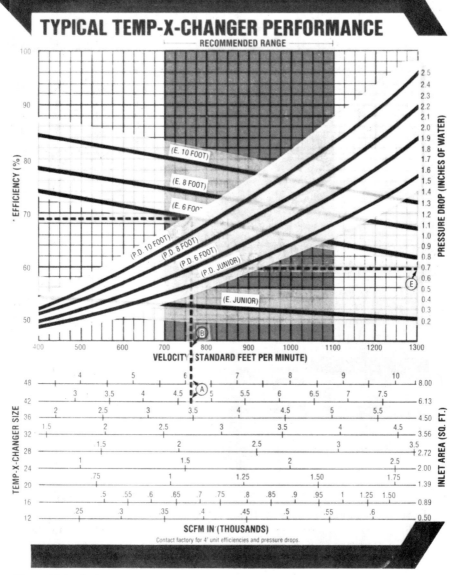

Figure 5.1-1. Gas/gas curved plate counter-flow heat exchanger sizing and performance chart. (*Courtesy United Air Specialists, Inc., Cincinnati, OH.*)

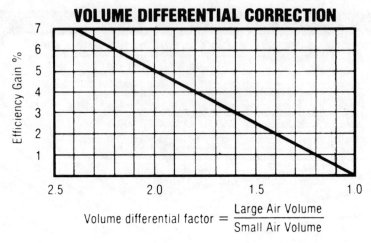

Figure 5.1-2. Gas/gas curved plate counter-flow heat exchanger volume differential correction. (*Courtesy United Air Specialists, Inc., Cincinnati, OH.*)

8. $\Delta T_e = \dfrac{1{,}409{,}319}{(1.08)(11{,}075)} = 118°\text{F}.$ (Equation 2-4)
9. $T_{el} = 210 - 118 = 92°\text{F}.$
10. Enter Figure 5.1-3 for $(92 + 210)/2 = 151°\text{F}$, find correction factor = 1.2. Exhaust pressure drop = 0.70 in. water.

Example 2: Equal flows of moist air.

$$V_{ee} = V_{se} = 16{,}700 \text{ cfm}$$

$$T_{ee} = 85° \text{ dry bulb; } 74°\text{ F wet bulb}$$

$$T_{se} = 10°\text{ F dry outside air}$$

1. $V_{\text{std},e} = \dfrac{(16{,}700)(530)}{(545)} = 16{,}240 \text{ standard cfm}.$ (Equation 2-5)
2. Select three units, 36 in. by 48 in. by 6 ft long.
3. Enter Figure 5.1-1 on 36 by 48 line at one-third $V_{\text{std},e}$ (5413.3 standard cfm), find $E = 67\%$. Find pressure drop on P.D. curve at 0.95 in. water.
4. $\Delta T = (0.67)(85\text{-}10) = 50°\text{ F}.$ (Equation 4-8)
5. $T_{si} = 10 + 50 = 60°\text{ F}.$
6. $W_e = (60)(16{,}240)(0.076) = 74{,}054 \text{ lb/hr}.$ (Equation 3-3)
7. $H = (1.08)(16{,}240)(50) = 876{,}960 \text{ Btu/hr}.$ (Equation 2-4)
8. $\Delta h = \dfrac{876{,}960}{74{,}054} = 11.8 \text{ Btu/lb}.$

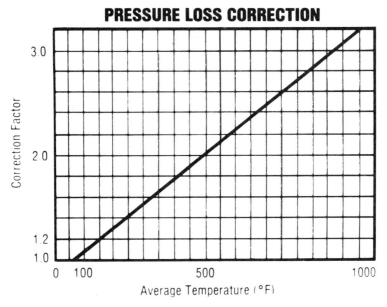

Figure 5.1-3. Gas/gas curved plate counter-flow heat exchanger pressure loss correction. (*Courtesy United Air Specialists, Inc., Cincinnati, OH.*)

9. Entering psychrometric chart, Figure 3.11-1, at given exhaust conditions, find $h_{ee} = 37.6$ Btu/lb.
10. $h_{el} = 37.6 - 11.8 = 25.8$ Btu/lb.
11. At exhaust entering conditions on Figure 3.11-1, follow horizontal line to saturation curve and along saturation curve to $h_{ee} = 25.8$. Find $T_{el} = 59°$ F, saturated. Condensation occurs in the exchanger.

5.2 GAS/GAS FLAT PLATE COUNTER-FLOW HEAT EXCHANGER

Example: Unequal air flow rates

$$V_e = 7000 \text{ cfm} \qquad T_{ee} = 95° \text{ F}$$
$$V_s = 9000 \text{ cfm} \qquad T_{se} = 0° \text{ F}$$

Desired $E = 75\%$. Pressure drop about 1 in. water.

1. $V_{\text{std}, e} = \dfrac{(7000)(530)}{(555)} = 6684$ standard cfm. \hfill (Equation 2-5)

   $V_{\text{std}, s} = \dfrac{(9000)(530)}{(460)} = 10{,}370$ standard cfm. \hfill (Equation 2-5)

144 INDUSTRIAL AND COMMERCIAL HEAT RECOVERY SYSTEMS

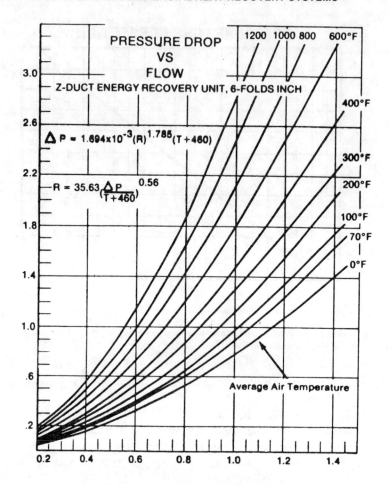

Figure 5.2-1. Gas/gas flat plate counter-flow heat exchanger pressure drop vs. flow rate. (*Courtesy Des Champs Laboratories, Inc., East Hanover, NJ.*)

2. $K = \dfrac{10,370}{6684} = 1.55$.
3. Enter Figure 5.2-1 at 1 in. water pressure drop and assume supply temperature within heat exchanger at 30°F, find $R = 1.11$.
4. Rated V_{std} of heat exchanger = $(10.370/1.11) = 9342$ standard cfm.
5. Enter Figures 5.2-2a and 5.2-2b, find 75M8 rated at 8000 standard cfm. R now equals $(10,370/8000) = 1.30$.

SIZING AND PERFORMANCE OF HEAT RECOVERY EQUIPMENT 145

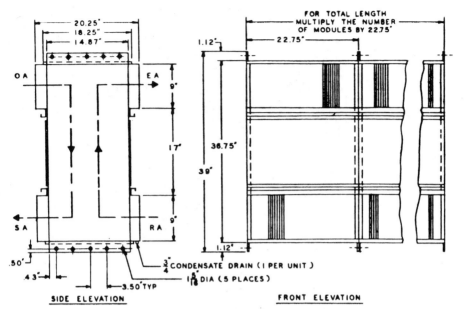

MODEL NO.	NOMINAL RATING CFM	DIMENSIONS (INCHES) L × W × H			OPERATING RANGE (CFM)	PRESSURE DROP (IN. WC)	EFFICIENCY (%)	NET WEIGHT (LBS)
74-1000-A6	1,000	39.00	17.50	22.50	500 .. 1,150	.4 ... 1.15	73 67	125
74-1000-AA6	1,000	39.00	17.50	22.50	500 .. 1,150	.4 ... 1.15	73 67	100
74-1000-SS5	1,000	39.00	17.50	22.50	500 .. 1,150	.4 ... 1.15	73 67	180

Figure 5.2-2a. Gas/gas flat plate counter-flow heat exchanger standard modules. (*Courtesy Des Champs Laboratories, Inc., East Hanover, NJ.*)

6. Enter Figure 5.2-1 and find $\Delta P_s = 1.32$ in. water.
7. Enter Figure 5.2-3 at $K = 1.55$ and $6684/8000 = 0.84$; find $E = 82\%$.
8. $T_{el} = 95 - (0.82)(95 - 0) = 17.1°$ F. (Equation 4-8)
9. $T_{sl} = 0 + (0.82)\left(\dfrac{1}{1.55}\right)(95 - 0) = 50.3°$ F. (Equation 4-8)
10. Average supply temperature within heat exchanger is $(0 + 50.3/2) = 25°$ F, close to assumption.
11. $H = (1.08)(6684)(95 - 17.1) = 562,338$ Btu/hr. (Equation 2-4)

No credit is taken for condensation in the heat exchanger. Frosting conditions can be checked by referring to Figure 5.2-4.

146 INDUSTRIAL AND COMMERCIAL HEAT RECOVERY SYSTEMS

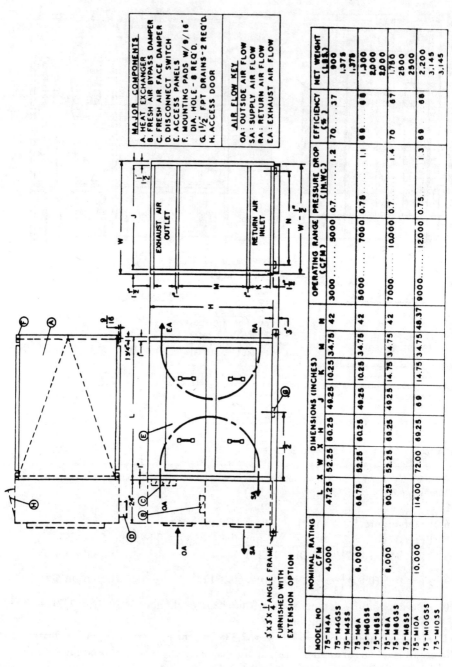

Figure 5.2-2b. Gas/gas flat plate counter-flow heat exchanger and standard modules. (*Courtesy Des Champs Laboratories, Inc., East Hanover, NJ.*)

SIZING AND PERFORMANCE OF HEAT RECOVERY EQUIPMENT

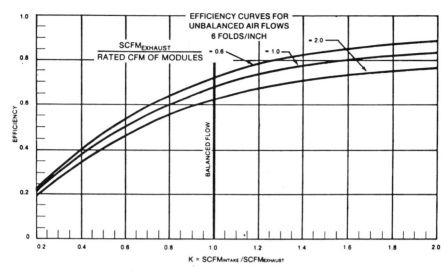

Figure 5.2-3. Gas/gas flat plate counter-flow heat exchanger efficiency curves. (*Courtesy Des Champs Laboratories, Inc., East Hanover, NJ.*)

FROST THRESHOLD TEMPERATURE, T_1, FOR VARIOUS EXHAUST AIR CONDITIONS

Exhaust Air Conditions			Ratio of Fresh Air Flow to Exhaust Air Flow, K							
T		RH%	0.5		0.7		1.0		2.0	
°C	°F		°C	°F	°C	°F	°C	°F	°C	°F
16	60	30	-16	2	-9	15	-5	23	0	32
16	60	40	-16	2	-9	15	-5	23	0	32
16	60	50	-20	-4	-13	9	-7	18	0	32
16	60	60	-22	-9	-15	4	-10	13	-5	32
21	70	30	-25	-13	-15	5	-8	17	-2	28
21	70	40	-29	-21	-19	-3	-12	10	-6	21
21	70	50	-33	-27	-22	-9	-16	3	-9	15
21	70	60	-35	-32	-25	-13	-18	-1	-12	10
24	75	30	-31	-25	-20	-4	-12	10	-5	23
24	75	40	-36	-33	-24	-12	-17	2	-9	15
24	75	50	-40	-40	-29	-20	-21	-6	-14	7
24	75	60	-44	-47	-32	-26	-24	-12	-17	1
27	80	30	-37	-35	-24	-11	-15	4	-8	19
27	80	40	-42	-44	-29	-20	-20	-5	-12	10
27	80	50	-47	-53	-34	-30	-25	-14	-17	1
27	80	60	-52	-62	39	-39	-30	-23	-22	-8
32	90	30	-50	-58	-34	-30	-24	-11	-15	5
32	90	40					-31	-24	-22	-8
32	90	50							-29	-20

Figure 5.2-4. Gas/gas flat plate counter-flow heat exchanger frost threshold temperature. (*Courtesy Des Champs Laboratories, Inc., East Hanover, NJ.*)

148 INDUSTRIAL AND COMMERCIAL HEAT RECOVERY SYSTEMS

5.3 GAS/GAS CROSS-FLOW HEAT EXCHANGERS

Example 1: Cross-flow heat exchanger from Figure 4.4-5. Equal flow rates of dry air.

$$V_{se} = V_{ee} = 5000 \text{ standard cfm} \quad T_{se} = 70°F$$
$$T_{ee} = 370°F$$

Desired $E = 60\%$.

1. $\Delta T = (0.6)(370 - 70) = 180°F$ (Equation 4-8)
   $T_{sl} = 70 + 180 = 250°F$
   $T_{el} = 370 - 180 = 190°F$
2. Enter Figure 5.3-2 at 60% effectiveness (efficiency). Move vertically to lower (efficiency) curve intersection, horizontally to left to ΔP curve, and vertically to 0.23 in. water, ΔP.
3. Correct for average temperature in heat exchangers.

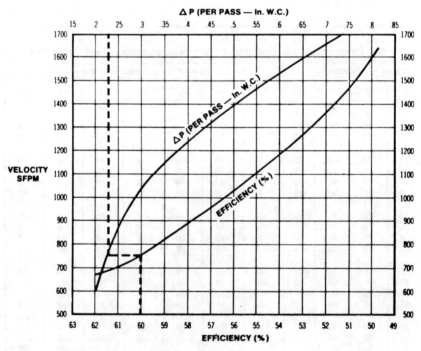

Figure 5.3-1. Cross-flow heat exchanger from Fig. 4.4-5 performance curves. (*Courtesy National-Standard Machinery Systems Division, Rome, NY.*)

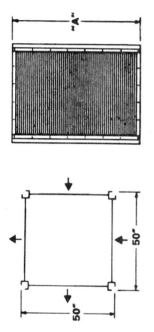

SIZES and WEIGHTS

SCFM	"A" in.	60% EFFICIENCY					"A" in.	55% EFFICIENCY					"A" in.	50% EFFICIENCY			
		Galvanized Wt. lbs.	Aluminized Steel Wt. lbs.	Aluminum Wt. lbs.	304 Stainless Wt. lbs.		Galvanized Wt. lbs.	Aluminized Steel Wt. lbs.	Aluminum Wt. lbs.	304 Stainless Wt. lbs.		Galvanized Wt. lbs.	Aluminized Steel Wt. lbs.	Aluminum Wt. lbs.	304 Stainless Wt. lbs.
1,000	14	795	845	355	775	11	540	580	245	530	9	385	412	175	375
2,000	24	1585	1685	705	1545	18	1077	1150	480	1050	14	770	818	343	750
3,000	34	2375	2525	1055	2315	25	1615	1725	720	1575	19	1150	1225	512	1120
4,000	44	3165	3365	1405	3085	32	2150	2295	955	2095	24	1530	1632	680	1490
5,000	54	3960	4205	1760	3860	39	2690	2875	1200	2625	29	1915	2033	855	1870
6,000	64	4750	5045	2110	4630	46	3225	3446	1435	3145	34	2300	2450	1025	2240
7,000	74	5540	5890	2460	5400	53	3760	4020	1675	3670	39	2675	2855	1190	2610
8,000	84	6327	6730	2810	6170	60	4300	4590	1910	4190	44	3055	3262	1360	2900
9,000	94	7115	7570	3160	6940	67	4835	5165	2150	4715	49	3440	3670	1530	3350
10,000	104	8010	8415	3515	7715	74	5375	5742	2390	5240	54	3825	4080	1700	3730
12,500	129	9890	10515	4395	9645	92	6710	7174	2990	6550	67	4785	5096	2135	4670
15,000	154	11870	12620	5275	11575	109	8065	8610	3590	7965	75	5735	6120	2560	5595

Figure 5.3-2. Cross-flow heat exchanger from Fig. 4.4-5 sizes and weights. (*Courtesy National-Standard Machinery Systems Division, Rome, N.Y.*)

150 INDUSTRIAL AND COMMERCIAL HEAT RECOVERY SYSTEMS

$$\text{Average supply temperature} = \frac{70 + 250}{2} = 160° \text{ F.}$$

$$\Delta P_s = \frac{(160 + 460)(0.23)}{(530)} = 0.27 \text{ in. water.}$$

$$\text{Average exhaust temperature} = \frac{370 + 190}{2} = 280° \text{ F.}$$

$$\Delta P_e = \frac{(280 + 460)(0.23)}{(530)} = 0.32 \text{ in. water.}$$

4. Enter Figure 5.3-3; find $A = 54$ in., weight = 4205 lb for aluminized steel.

Example 2: Paper cross-flow heat exchanger. Equal flow rates of moist air.

$$V_{ee} = V_{se} = 2943 \text{ standard cfm}$$

$$T_{se} = 19° \text{ F} \qquad T_{ee} = 72° \text{ F}$$

$$\text{Humidity} = 90\% \qquad \text{Humidity} = 36\%$$

1. Enter psychrometric chart, Figure 3.11-1; find:

$$w_{se} = 13 \text{ gr/lb dry air}$$
$$w_{ee} = 42 \text{ gr/lb dry air}$$
$$h_{se} = 7 \text{ Btu/lb}$$
$$h_{ee} = 24 \text{ Btu/lb}$$

2. Enter Figure 5.3-4; find:

$$E_{\text{sensible}} = 72\%$$
$$E_{\text{latent}} = 67.4\%$$
$$\Delta P = 1.2 \text{ in. water}$$

3. $w_{sl} = 13 + (0.674)(42 - 13) = 32.5$ gr/lb. (Equation 4-8)
   $w_{el} = 42 - (0.674)(42 - 13) = 22.3$ gr/lb. (Equation 4-8)
4. $T_{sl} = 19 + (0.72)(72 - 19) = 57.2°$ F. (Equation 4-8)
   $T_{el} = 72 - (0.72)(72 - 19) = 33.8°$ F. (Equation 4-8)
5. From psychrometric chart, Figure 3.11-1, find:

$$h_{sl} = 19 \text{ Btu/lb dry air.}$$
$$h_{el} = 12 \text{ Btu/lb dry air.}$$

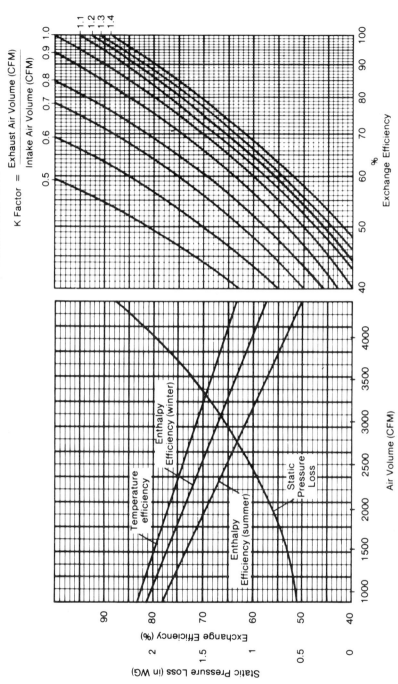

Figure 5.3-3. Paper cross-flow heat exchanger performance curve. (*Courtesy Mitsubishi Electric Sales America, Inc., Crompton, CA.*)

151

Outside Dimensions (in)

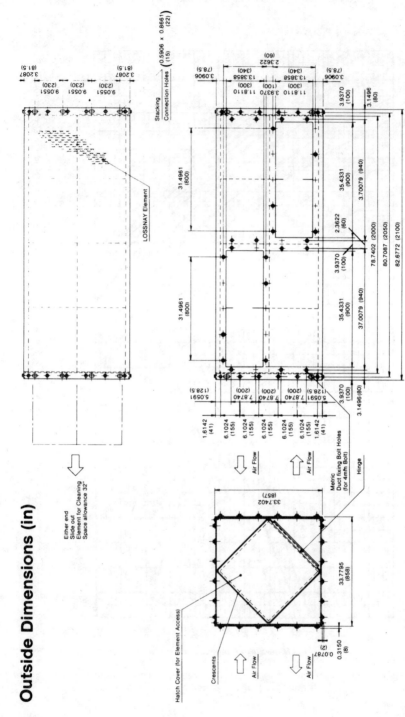

Figure 5.3.4. Paper cross-flow heat exchanger dimensional drawing. (*Courtesy Mitsubishi Electric Sales America, Inc., Compton, CA.*)

SIZING AND PERFORMANCE OF HEAT RECOVERY EQUIPMENT 153

6. $H_s = (2943)(0.076)(60)(19 - 7) = 161{,}041$ Btu/hr. (Equation 3-10)
  $H_e = (2943)(0.076)(60)(24 - 12) = 161{,}041$ Btu/hr. (Equation 3-10)

A dimensional drawing of the heat exchanger is shown in Figure 5.3-5. The manufacturer also supplies other models of similar construction, ranging from 54-cfm to 7063-cfm capacity.

5.4 GAS/GAS SHELL-AND-TUBE HEAT EXCHANGER

Example: Single-pass heat exchanger.

$$V_{se} = V_{ee} = 2628 \text{ standard cfm}$$
$$T_{se} = 70°\text{F}$$
$$T_{ee} = 750°\text{F}$$

Desired: Maximum T_{sl}

1. Enter Figure 5.4-1; assume HD-32; find semi-section area of 2.79 ft.^{2}

$$v = \frac{2628}{2.79} = 942 \text{ fpm} \qquad \text{(Equation 3-2)}$$

2. Enter Figure 5.4-2; find $E = 51\%$ and $\Delta P_s = 1.0$ in. water, $\Delta P_e = 0.8$ in. water. Several trials can be made of other sizes.
3. $\Delta T = (0.51)(750 - 70) = 347°$ F. (Equation 4-8)
4. $H = (1.08)(2628)(347) = 98{,}487$ Btu/hr. (Equation 2-4)

FOR UNIT CAPACITY: Multiply velocity (feet per minute) by the factor to obtain cubic feet per minute.			
HD-12	.39	HD-48	6.28
HD-16	.70	HD-54	7.95
HD-20	1.09	HD-60	9.82
HD-24	1.57	HD-66	11.88
HD-28	2.14	HD-72	14.14
HD-32	2.79	HD-78	16.60
HD-36	3.53	HD-84	19.24
HD-42	4.81	HD-90	20.09
		HD-96	25.13

Figure 5.4-1. Single-pass shell-and-tube heat exchanger semi-section areas. (*Courtesy United Air Specialists, Inc., Cincinnati, OH.*)

154 INDUSTRIAL AND COMMERCIAL HEAT RECOVERY SYSTEMS

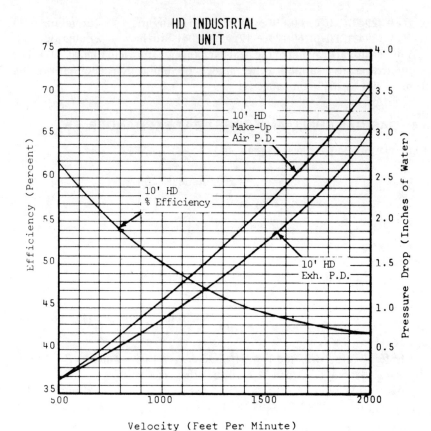

Figure 5.4-2. Single-pass shell-and-tube heat exchanger performance curves. (*Courtesy United Air Specialists, Inc., Cincinnati, OH.*)

5. If V_{se} and V_{ee} are unequal, calculate ratio of larger to smaller, enter Figure 5.4-3 to find effectiveness (efficiency) gain to be added to smaller flow, and calculate H based on smaller flow. Calculate larger flow leaving temperature from Equation 2-4.
6. $T_{el} = 750 - 347 = 403°\,\text{F}$.
7. Average T_e in exchanger $= \dfrac{750 + 403}{2} = 577°\,\text{F}$.
8. Average V_e in exchanger $= (2628)\left(\dfrac{460 + 577}{460 + 70}\right) = 5142$ cfm.
9. Average v_e in exchanger $= \dfrac{5142}{2.79} = 1843$ fpm.
10. Enter Figure 5.4-2; find $\Delta P_s = 3.0$ in. water, and $\Delta P_e = 2.5$ in. water.

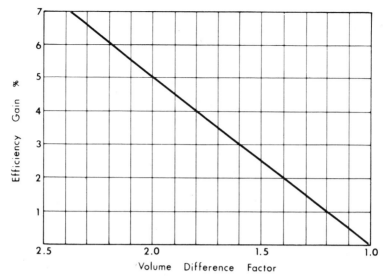

Figure 5.4-3. Single-pass shell-and-tube heat exchanger volume differential correction. (*Courtesy United Air Specialists, Inc., Cincinnati, OH.*)

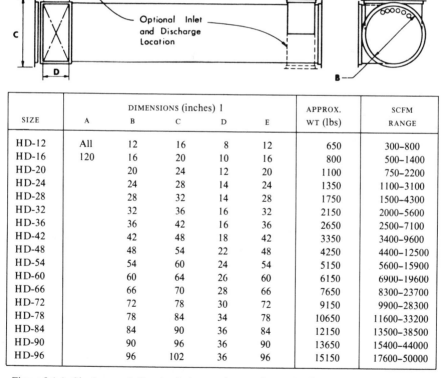

SIZE	DIMENSIONS (inches) 1					APPROX. WT (lbs)	SCFM RANGE
	A	B	C	D	E		
HD-12	All	12	16	8	12	650	300–800
HD-16	120	16	20	10	16	800	500–1400
HD-20		20	24	12	20	1100	750–2200
HD-24		24	28	14	24	1350	1100–3100
HD-28		28	32	14	28	1750	1500–4300
HD-32		32	36	16	32	2150	2000–5600
HD-36		36	42	16	36	2650	2500–7100
HD-42		42	48	18	42	3350	3400–9600
HD-48		48	54	22	48	4250	4400–12500
HD-54		54	60	24	54	5150	5600–15900
HD-60		60	64	26	60	6150	6900–19600
HD-66		66	70	28	66	7650	8300–23700
HD-72		72	78	30	72	9150	9900–28300
HD-78		78	84	34	78	10650	11600–33200
HD-84		84	90	36	84	12150	13500–38500
HD-90		90	96	36	90	13650	15400–44000
HD-96		96	102	36	96	15150	17600–50000

Figure 5.4-4. Single-pass shell-and-tube heat exchanger sizes and weights. (*Courtesy United Air Specialists, Inc., Cincinnati, OH.*)

156 INDUSTRIAL AND COMMERCIAL HEAT RECOVERY SYSTEMS

11. Correct for temperature, $\Delta P_s = (3.0) \dfrac{(460 + 70)}{(460 + 577)} = 1.5$ in. water.

$$\Delta P_e = (2.5) \dfrac{(460 + 70)}{(460 + 577)} = 1.3 \text{ in. water.}$$

12. Find dimensions in Figure 5.4-4.

5.5 GAS/GAS HEAT PIPE HEAT EXCHANGER

Example: Unequal flows of clean, dry air.

$$V_{se} = 6009 \text{ cfm} \qquad T_{se} = -5°\text{F}$$

$$V_{ee} = 12{,}000 \text{ cfm} \qquad T_{ee} = 300°\text{F}$$

1. $V_{\text{std}, s} = \dfrac{(6009)(530)}{(460 - 5)} = 7000$ standard cfm. (Equation 2-5)

   $V_{\text{std}, e} = \dfrac{(12{,}000)(530)}{(460 + 300)} = 8370$ standard cfm. (Equation 2-5)

2. Assume face velocity of 500 fpm.

$$A = \dfrac{7000 + 8370}{500} = 30.7 \text{ ft}^2. \qquad \text{(Equation 3-2)}$$

3. Enter Figure 5.5-1, find AP/CP 40.5 in. high by 108 in. wide, area 30.4 ft², flow area equal for supply and exhaust.

4. $v_{se} = \dfrac{7000}{(30.4/2)} = 461$ fpm. (Equation 3-2)

   $v_{ee} = \dfrac{8370}{(30.4/2)} = 551$ fpm. (Equation 3-2)

   Average entering velocity,

$$v = \dfrac{461 + 551}{2} = 506 \text{ fpm.}$$

5. $M = \dfrac{8370}{7000} = 1.20.$

6. Assume 8 rows, 8 fins per in. for exhaust, 14 fins per in. for supply, enter Figure 5.5-2, and find $E(R_a) = 65\%$ or 0.65. For applications with contamination, reduce fin spacing to 8 fins per in. or 5 fins per in. for extreme cases (use Figure 5.5-3). Use Figure 5.5-2 for aluminum or copper

SIZING AND PERFORMANCE OF HEAT RECOVERY EQUIPMENT   157

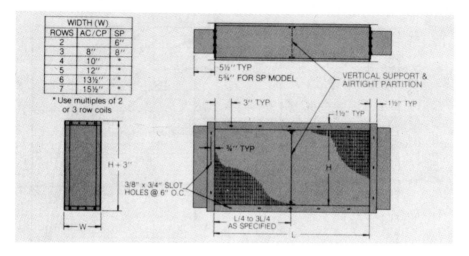

TRU DIMENSIONS

Dimension L	UNIT DIMENSIONS AND TOTAL FACE AREA - SQ. FT.												
	Dimension H												
	AP/CP	AP/CP	AP/CP	AP/CP	AP/CP	AP/CP	AP/CP	SP	SP	SP	SP	SP	SP
	13.5"	20.3"	27"	33.8"	40.5"	47.3"	54"	13"	21"	27"	33"	41"	47"	53"
24"	2.3	3.4						2.2	3.5				
36"	3.4	5.1	6.8	8.5				3.3	5.3	6.8	8.3		
48"	4.5	6.8	9.0	11.3	13.5	15.8		4.3	7.0	9.0	11.0	13.7	15.7
60"	5.6	8.5	11.3	14.1	16.9	19.7	22.5	5.4	8.8	11.3	13.8	17.1	19.6	22.1
72"	6.8	10.2	13.5	16.9	20.3	23.7	27.0	6.5	10.5	13.5	16.5	20.5	23.5	26.5
84"		11.8	15.8	19.7	23.6	27.6	31.5		12.3	15.8	19.3	23.9	27.4	30.9
96"		13.5	18.0	22.5	27.0	31.5	36.0		14.0	18.0	22.0	27.3	31.3	35.3
108"		15.2	20.3	25.4	30.4	35.5	40.5		15.8	20.3	24.8	30.8	35.3	39.8
120"			22.5	28.2	33.8	39.4	45.0			22.5	27.5	34.2	39.2	44.2
132"			24.8	31.0	37.1	43.4	49.5			24.8	30.3	37.6	43.1	48.6
144"			27.0	33.8	40.5	47.3	54.0			27.0	33.0	41.0	47.0	53.0
156"			29.3	36.6	43.9	51.2	58.5							
168"				39.4	47.3	55.2	63.0							
180"				42.3	50.6	59.1	67.5							
192"				45.1	54.0	63.1	72.0							

NOTE: MAXIMUM AVAILABLE LENGTH ON TYPE SP UNITS IS 144"

Figure 5.5-1. Heat pipe heat exchanger sizes and dimensions. (*Courtesy Q-dot Corp., Dallas, TX.*)

construction and T_e below 525° F. Use Figure 5.5-3 for steel construction and T_e below 825° F.

7. For largest flow rate,

$$T_{el} = 300 - \frac{(0.65)(300 + 5)}{1.20} = 135° \text{F}. \quad \text{(Equation 4-8)}$$

For smallest flow rate,

$$T_{sl} = -5 + (0.65)(300 + 5) = 193° \text{F}. \quad \text{(Equation 4-8)}$$

158 INDUSTRIAL AND COMMERCIAL HEAT RECOVERY SYSTEMS

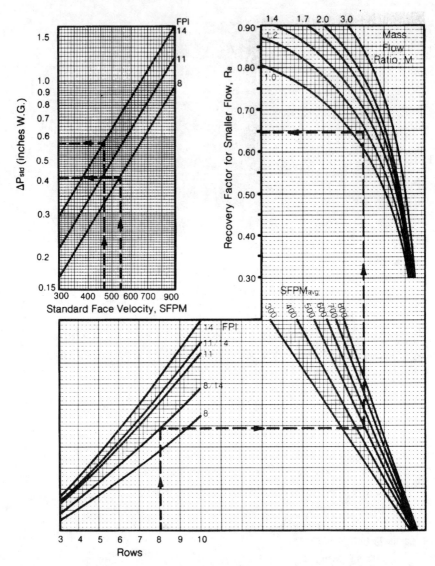

Figure 5.5-2. Heat pipe exchanger performance curves for copper and aluminum heat pipes. (*Courtesy Q-dot Corp., Dallas, TX.*)

SIZING AND PERFORMANCE OF HEAT RECOVERY EQUIPMENT   159

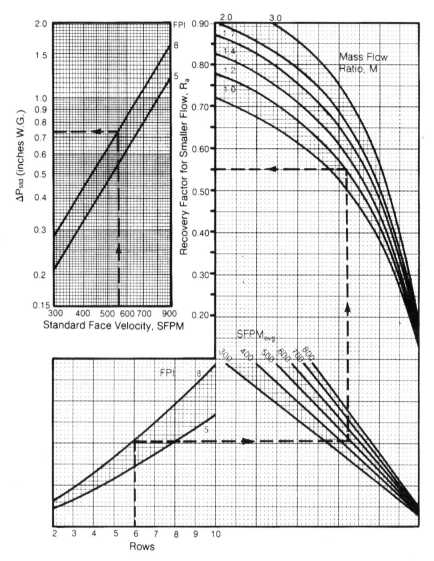

Figure 5.5-3. Heat pipe exchanger performance curves for steel heat pipes. (*Courtesy Q-dot Corp., Dallas, TX.*)

160 INDUSTRIAL AND COMMERCIAL HEAT RECOVERY SYSTEMS

8. $H = (1.08)(7000)(193 + 5) = 1,496,880$ Btu/hr. (Equation 2-4)
9. Enter Figure 5.5-2 at v_{se} and 14 fins per in.; find $\Delta P_s = 0.57$ in. water.
   Enter Figure 5.5-2 at v_{ee} and 8 fins per in.; find $\Delta P_e = 0.41$ in. water.
10. Average heat exchanger $T_s = \dfrac{-5 + 193}{2} = 99°F$.

    Average heat exchanger $T_e = \dfrac{300 + 135}{2} = 218°F$.
11. Figure 5.5-2 is ΔP for five rows. Correct for eight rows and average exchanger temperature.
12. $\Delta P_s = (0.57)\left(\dfrac{8}{5}\right)\left(\dfrac{460 + 99}{530}\right) = 0.96$ in. water.

    $\Delta P_e = (0.41)\left(\dfrac{8}{5}\right)\left(\dfrac{460 + 218}{530}\right) = 0.84$ in. water.
13. Repeat the calculations with other sizes if ΔP is not acceptable.
14. $Z = \dfrac{(12)(1,496,880)}{(8)(40.5)} = 55,440$.

    Heat pipe can be installed at any angle to vertical if Z is less than 253,000.
15. Consult manufacturer for possible frosting conditions.

5.6 HEAT WHEELS

Example 1: Moisture—transfer heat wheel. Unequal volumes of moist air.

$V_s = 7760$ standard cfm $T_s = 90°F$ dry bulb; $76°F$ wet bulb.

$V_e = 9760$ standard cfm $T_e = 75°F$ dry bulb; 50% relative humidity.

1. Enter Figure 5.6-1, at larger flow and select WEC − 1400 wheel. Find face velocity = 700 fpm and $\Delta P_e = 0.77$ in. water.
2. Enter Figure 5.6-1 at smaller flow and same wheel. Face velocity interpolated as:

$$500 + \dfrac{(600 - 500)(7760 - 6960)}{(8350 - 6960)} = 557 \text{ fpm}$$

3. $\Delta P_s = 0.55 + \dfrac{(0.66 - 0.55)(557 - 500)}{(600 - 500)} = 0.61$ in. water.
4. $E_s = 78.5 - \dfrac{(78.5 - 75.5)(557 - 500)}{(600 - 500)} = 76.8\%$.
5. $K = \dfrac{9700}{7760} = 1.25$.

SIZING AND PERFORMANCE OF HEAT RECOVERY EQUIPMENT 161

Model No.	FLOW RATE — SCFM									
WEC-350	660	990	1320	1650	1970	2300	2630	2960	3300	3620
WEC-600	1050	1581	2110	2640	3160	3690	4220	4740	5270	5800
WEC-875	1620	2440	3250	4060	4870	5680	6500	7310	8120	8930
WEC-1400	2780	4170	5564	6960	8350	9740	11100	12500	13900	15300
WEC-1700	3530	5300	7070	8840	10600	12400	14100	15900	17700	19400
WEC-2150	4450	6670	8900	11100	13300	15600	17800	20000	22200	24500
WEC-3160	6550	9830	13100	16400	19650	22900	26200	29500	32750	36000
WEC-4770	9660	14500	19300	24150	29000	33800	38600	43500	48300	53100

VELOCITY—FPM										
200	300	400	500	600	700	800	900	1000	1100

BASE EFFICIENCY % (Based on Average Wheel Temperature of 70°F.) (±3%)										
85.5	83.5	81	78.5	75.5	73	70.5	67.5	66	64.5

PRESSURE DROP — INS. W. C. (70°F.)										
.22	.34	.45	.55	.66	.77	.89	1.01	1.15	1.33

Figure 5.6-1. Moisture-transfer heat wheel performance chart. (*Courtesy The Wing Co., Cranford, NJ.*)

162 INDUSTRIAL AND COMMERCIAL HEAT RECOVERY SYSTEMS

	CORRECTED EFFICIENCY UNEQUAL AIR FLOW						
	BASE EFFICIENCY						
K*	60	65	70	75	80	85	90
.5	34	38	41	44	46	48	50
.6	40	44	48	51	54	57	59
.7	46	50	54	58	62	66	68
.8	51	55	60	64	69	73	77
.9	.55	60	65	70	75	79	84
1.0	60	65	70	75	80	85	90
1.1	61	67	72	77	83	88	93
1.2	63	68	74	79	85	90	95
1.3	64	70	75	81	87	92	97
1.4	65	71	77	82	88	93	98
1.5	66	72	78	84	89	94	99
1.6	67	73	79	85	90	95	99
1.7	67	73	80	85	91	96	99
1.8	68	74	80	86	92	96	99
1.9	68	75	81	87	92	97	99
2.0	69	75	81	87	93	97	99

*K Unequal air flow ratio of $\frac{\text{Exhaust SCFM}}{\text{Supply SCFM}}$

Figure 5.6-2. Moisture-transfer heat wheel unequal air flow correction. (*Courtesy The Wing Co., Cranford, NJ.*)

6. Enter Figure 5.6-2; find corrected E_S by interpolation.

$$E_S = 80 + \frac{(86 - 80)(76.8 - 75)}{(80 - 75)} = 82.2\%.$$

Sensible and latent effectivenesses are equal at 18 rpm wheel speed used in Figures 5.6-1 and 5.6-2.

7. $T_{sl} = 90 - (0.822)(90 - 75) = 77.7°F.$ (Equation 4-8)
8. Enter psychrometric chart, Figure 3.11-1; find:

$$w_{se} = 113 \text{ gr/lb dry air.}$$

$$w_{ee} = 65 \text{ gr/lb dry air.}$$

SIZING AND PERFORMANCE OF HEAT RECOVERY EQUIPMENT 163

9. $w_{sl} = 113 - (0.822)(113 - 65) = 73.5$ gr/lb dry air. (Equation 4-8)
10. Enter psychrometric chart, Figure 3.11-1; find:

$$h_{se} = 39.3 \text{ Btu/lb dry air.}$$

$$h_{sl} = 29.8 \text{ Btu/lb dry air.}$$

11. $H = (7760)(0.076)(60)(39.3 - 29.8) = 336,163$ Btu/hr. (Equation 3-10)
12. Check frosting conditions by locating exhaust entering and leaving conditions on psychrometric chart, Figure 3.11-1, and connecting with a straight line. If line crosses saturation curve, frosting will occur. Pre-heat supply entering air enough to avoid crossing saturation curve.
13. Wheel dimensions are shown in Figure 5.6-3.

Example 2: Metal heat wheel. Equal volumes of dry air.

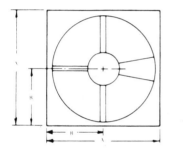

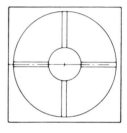

Building Side   Weather Side

Model No.	A	B	C	Face Area Per Side	Net Weight Lbs.
350	42	21	12	3.3	490
600	53½	26¾	15⅜	5.27	710
875	64	32	15⅜	8.12	920
1400	83	41½	16¼	13.9	1520
1700	91	45½	16⅝	17.7	2100
2150	101	50½	16⅝	22.2	2650
3160	120	60	19⅝	32.8	3150
4770	144	72	19⅝	48.3	4150

Figure 5.6-3. Moisture-transfer heat wheel dimensions. (*Courtesy The Wing Co., Cranford, NJ.*)

Model No.	FLOW RATE — SCFM												
WCM-125	220	330	440	560	670	780	890	1000	1100	1220	1330	1440	1550
WCM-350	590	890	1180	1480	1780	2070	2370	2660	2960	3260	3550	3850	4140
WCM-600	950	1425	1900	2380	2850	3330	3800	4280	4750	5230	5700	6180	6650
WCM-875	1490	2238	2980	3730	4480	5220	5970	6710	7460	8210	8950	9700	10400
WCM-1400	2580	3870	5160	6460	7750	9040	10300	11600	12900	14200	15500	16800	18100
WCM-2150	4090	6140	8180	10200	12300	14300	16400	18400	20950	22500	24500	26600	28600
WCM-3160	6120	9180	12200	15300	18400	21400	24500	27500	30600	33700	36700	39800	42800

VELOCITY — FPM

200	300	400	500	600	700	800	900	1000	1100	1200	1300	1400

EFFICIENCY % (Based on Average Wheel Temperature of 300° F.) (±3%)

84	82	79	76	73	70	68	65	63	61	60	58	57

PRESSURE DROP — INS. W. C. (70° F.)

.10	.16	.22	.29	.35	.42	.48	.56	.62	.69	.77	.84	.91

Figure 5.6-4. Metal heat wheel performance chart. (*Courtesy The Wing Co., Cranford, NJ.*)

SIZING AND PERFORMANCE OF HEAT RECOVERY EQUIPMENT

Avg. Temp. °F.	BASE EFF. (AVG. TEMP. = 300°F.)						
	55	60	65	70	75	80	85
	TEMP. CORRECTED EFFICIENCIES						
150	51	56	61	66	72	77	82
200	52	57	62	67	73	78	83
250	54	59	64	69	74	79	84
300	55	60	65	70	75	80	85
350	56	61	66	71	76	81	86
400	58	63	67	72	78	81	86
450	59	64	68	73	78	82	87

Figure 5.6-5. Metal heat wheel temperature correction chart. (*Courtesy The Wing Co., Cranford, NJ.*)

$$V_s = V_e = 10,000 \text{ cfm} \qquad T_s = 0°\text{F}$$
$$T_e = 500°\text{F}$$

1. $V_{\text{std}, e} = \dfrac{(10,000)(530)}{(960)} = 5520$ standard cfm. (Equation 2-5)
2. Average wheel $T = \dfrac{500 + 0}{2} = 250°$ F.
3. Enter Figure 5.6-4, select WCM − 875, and find $E = 69\%$.
4. Enter Figure 5.6-5; find temperature—corrected $E = 68\%$.
5. If flows are unequal, use Figure 5.6-2.
6. $T_{sl} = 0 + (0.68)(500 − 0) = 340°$ F. (Equation 4-8)
   $T_{el} = 500 − (0.68)(500 − 0) = 160°$ F. (Equation 4-8)
7. $H = (1.08)(5520)(340 − 0) = 2,026,944$ Btu/hr. (Equation 2-4)
8. Check frosting conditions, as in Example 1.
9. Wheel dimensions are shown in Figure 5.6-6.

5.7 GAS/GAS HIGH TEMPERATURE HEAT EXCHANGER

Example 1: High temperature cross-flow heat exchanger. Equal flow rates of supply air and combustion products.

$$V_s = 5500 \text{ standard cfh–air; } 500 \text{ standard cfh natural gas}$$
$$V_e = 6000 \text{ standard cfh}$$
$$T_s = 70°\text{F}$$
$$T_e = 2000°\text{F}$$

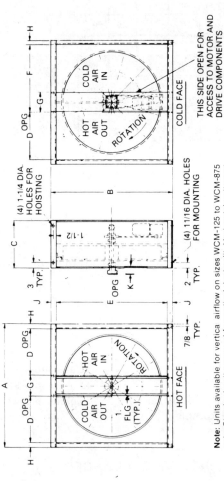

Figure 5.6-6. Metal heat wheel dimensions. (*Courtesy: The Wing Co., Cranford, N.J.*)

SIZING AND PERFORMANCE OF HEAT RECOVERY EQUIPMENT   167

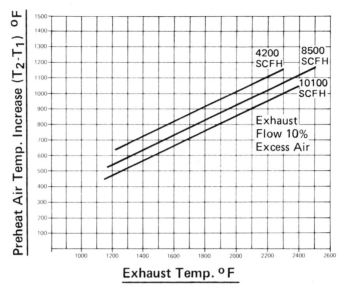

Figure 5.7-1. High-temperature cross-flow heat exchanger Model R 0600 TPX performance curves. (*Courtesy GTE Products Corp., Towanda, PA.*)

1. Enter Figure 5.7-1; for Model R0600 TPX find $\Delta T = 950°$ F.
2. $H = \dfrac{(1.08)(6000)(950)}{(60)} = 102{,}600$ Btu/hr.   (Equation 2-4)
3. Enter Figure 5.7-2; for Model R0600 TPX, find $\Delta P_s = 18.0$ in. water and $\Delta P_e = 0.3$ in. water. Enter Figure 5.7-6 for dimensions.
4. $T_{sl} = 70 + 950 = 1020°$ F
   $T_{el} = 2000 - 950 = 1050°$ F.
5. $E = \dfrac{(950)(100)}{(2000 - 70)} = 49.2\%$.   (Equation 4-8)
6. Larger flows are sized using the R1000 TPX size shown in Figures 5.7-3, 5.7-4, and 5.7-5. A larger model, handling flows to 100,000 standard cfh, is also available.

Example 2: Counter-flow radiant tube heat exchanger. Equal flow rates of supply air and combustion products.

$$V_S = V_e = 10{,}000 \text{ standard cfh} \qquad T_{se} = 70° \text{ F}$$
$$T_{ee} = 1600° \text{ F}$$

1. Enter Figure 5.7-7; find $T_{sl} = 800°$ F.

168 INDUSTRIAL AND COMMERCIAL HEAT RECOVERY SYSTEMS

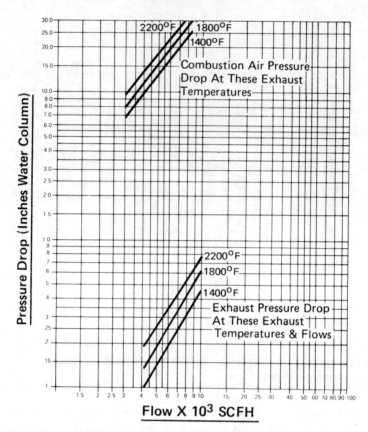

Figure 5.7-2. High-temperature cross-flow heat exchanger Model R 0600 TPX pressure drop curves. (*Courtesy GTE Products Corp., Towanda, PA.*)

2. $H = \dfrac{(1.08)(10,000)(800 - 70)}{(60)} = 131{,}400$ Btu/hr.   (Equation 2-4)
3. Enter Figure 5.7-8 for model EM 150; find $\Delta P_s = 10$ in. water.
4. $E = \dfrac{(800 - 70)(100)}{(1600 - 70)} = 47.7\%$.   (Equation 4-8)
5. Find dimensions on Figure 5.7-9.

Example 3: Single-pass radiant tube heat exchanger. Equal flow of supply air, and combustion products.

$$T_{se} = 70°\text{ F}$$

Heat input = 3,500,000 Btu/hr natural gas.

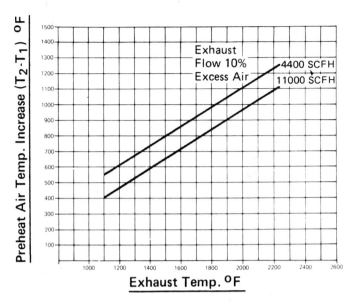

Figure 5.7-3. High-temperature cross-flow heat exchanger Model R 1000 TPX performance curves. (*Courtesy GTE Products Corp., Towanda, PA.*)

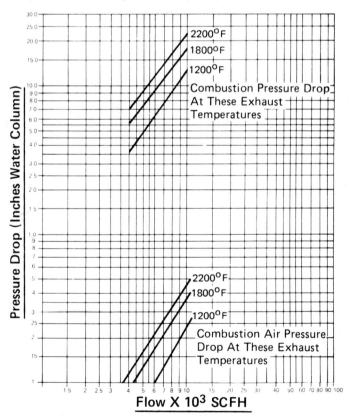

Figure 5.7-4. High-temperature cross-flow heat exchanger Model R 1000 TPX pressure drop curves. (*Courtesy GTE Products Corp., Towanda, PA.*)

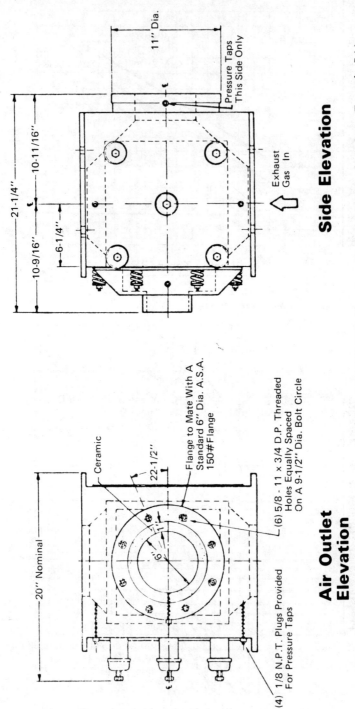

Figure 5.7-5. High-temperature cross-flow heat exchanger Model R 1000 TPX dimensions. (*Courtesy GTE Products Corp., Towanda, PA.*)

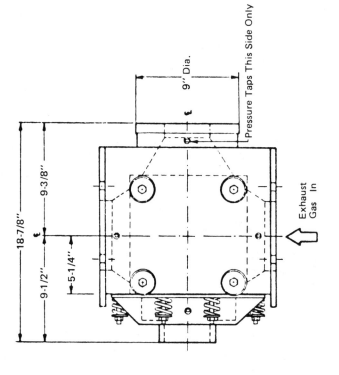

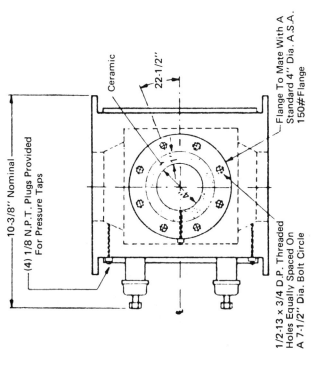

Figure 5.7-6. High-temperature cross-flow heat exchanger model R0600 TPX dimensions. (*Courtesy GTE Products Corp., Towanda, PA.*)

172 INDUSTRIAL AND COMMERCIAL HEAT RECOVERY SYSTEMS

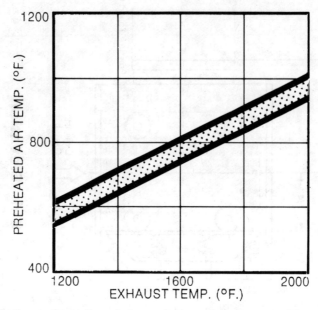

Figure 5.7-7. Counter-flow radiant tube heat exchanger performance curves. (*Courtesy Holcroft Div., Livonia, MI.*)

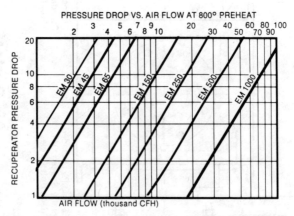

Figure 5.7-8. Counter-flow radiant tube heat exchanger pressure drop curves. (*Courtesy Holcroft Div., Livonia, MI.*)

SIZING AND PERFORMANCE OF HEAT RECOVERY EQUIPMENT

APPLICATION DATA	MODEL NUMBERS						
	EM 30	EM 45	EM 65	EM 150	EM 250	EM 500	EM 1000
Air Capacity, cfh*	3,000	4,500	6,500	15,000	25,000	50,000	100,000
Inner diameter, inches	6	9	10	14	18	24	26
Overall Height, Inches	49¼	49¼	49¼	61¼	91¼	121⅜	133⅜
Air Inlet, pipe size	1½	1½	2½	3	4	6†	8†

Figure 5.7-9. Counter-flow radiant heat exchanger dimensions. (*Courtesy Holcroft Div., Livonia, MI.*)

1. Enter Figure 5.7-10, Graph 1; choose highest available $T_{sl} = 820°$ F for size DD.
2. Enter Figure 5.7-10, Graph 2; find $\Delta P_s = 4.6$ in. water for size DD.
3. If ΔP_s is unacceptable, try other sizes.
4. Find dimensions on unit size chart.
5. $V_{std, s} = \dfrac{3,500,000}{10} = 35,000$ standard cfh.
6. $H = \dfrac{(1.08)(35,000)(820 - 70)}{(60)} = 472,500$ Btu/hr. (Equation 2-4)
7. $E = \dfrac{(750)(100)}{(2100 - 70)} = 36.9\%$. (Equation 4-8)

5.8 COMBUSTION AIR PRE-HEAT

Example 1: Single-pass radiant tube heat exchanger. From Example 3, Section 5.7:

$$T_{ee} = 2100° \text{ F}$$
$$T_{sl} = 820° \text{ F}$$
$$\text{Heat input} = 3,500,000 \text{ Btu/hr.}$$

Fuel–natural gas, 10% excess air.

1. Enter figure 2.4-1, find F by interpolation:

$$F = 27.2 + \dfrac{(820 - 800)(29.9 - 27.2)}{(900 - 800)} = 27.7\%.$$

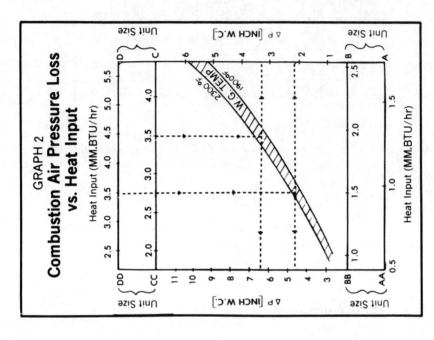

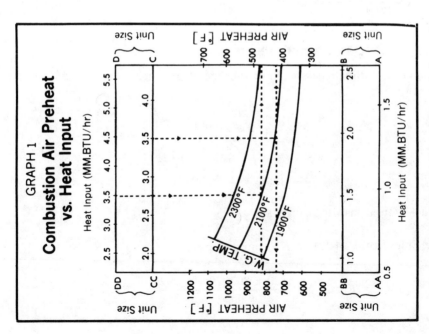

Figure 5.7-10. Single-pass radiant tube heat exchanger. (*Courtesy Thermal Transfer Corp., Monroeville, PA.*)

SIZING AND PERFORMANCE OF HEAT RECOVERY EQUIPMENT 175

2. $H = \dfrac{(27.7)(3{,}500{,}000)}{(100)} = 969{,}500$ Btu/hr.   (Equation 2-7)

3. Gas saved $= \dfrac{969{,}500}{1020} = 950.5$ cfh natural gas. Natural gas = 1020 Btu/ft^3.

Example 2: Burner with triple-pass integral heat exchanger. Fuel—natural gas, 10% excess air.

$$T_{ee} = 2200° \text{F}$$

$$\text{Heat input} = 1{,}000{,}000 \text{ Btu/hr.}$$

1. Enter Figure 5.8-1, find $T_{sl} = 1170°$ F.
2. Enter Figure 2.4-1; find F by interpolation:

$$F = 36.5 + \dfrac{(1170 - 1100)(38.8 - 36.5)}{(1200 - 1100)} = 38.1\%.$$

3. $H = \dfrac{(38.1)(1{,}000{,}000)}{(100)} = 381{,}000$ Btu/hr.   (Equation 2-7)

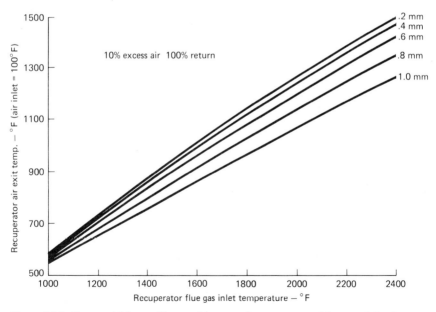

Figure 5.8-1. Burner with integral heat exchanger performance curve. (*Courtesy Selas Corporation of America. Dresher, PA.*)

176 INDUSTRIAL AND COMMERCIAL HEAT RECOVERY SYSTEMS

4. Gas saved $\dfrac{381{,}000}{1020} = 373.5$ cfh natural gas.

 Natural gas = 1020 Btu/ft^3.

Example 3: Self-recuperative burner.
 Fuel—natural gas, 10% excess air.

$$T_{ee} = 2200°\text{ F}$$

Heat input = 1,000,000 Btu/hr.

1. Enter Figure 5.8-2; find $T_{sl} = 810°$ F.
2. Enter Figure 2.4-1; find F by interpolation:

$$F = 27.2 + \frac{(810 - 800)(29.9 - 27.2)}{(900 - 800)} = 27.5\%.$$

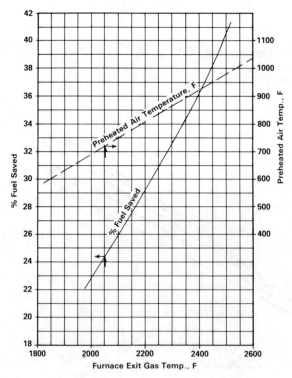

Figure 5.8-2. Self-recuperative burner performance curve. (*Courtesy North American Mfg. Co., Cleveland, OH.*)

SIZING AND PERFORMANCE OF HEAT RECOVERY EQUIPMENT 177

3. $H = \dfrac{(27.5)(1,000,000)}{(100)} = 275,000$ Btu/hr. (Equation 2-7)

4. Gas saved $= \dfrac{275,000}{1020} = 269.6$ cfh natural gas.

Burner dimensions are available from manufacturers.

5.9 GAS/LIQUID HEAT EXCHANGERS

Example 1: Clean air heated by hot waste water.

$$V_{se} = 6000 \text{ standard cfm} \qquad T_{se} = 60°\text{F}$$
$$T_{ee} = 200°\text{F}$$

$$\text{Desired } T_{sl} = 150°\text{F}.$$

$$T_{el} = 180°\text{F}.$$

Select coil for maximum heat transfer.
1. Assume coil face velocity = 500 fpm.
   Coil face area = (6000/500) = 12 ft². (Equation 3-2)
2. Enter Figure 5.9-1; find 24 in. by 72 in. coil with 12 ft² face area.
3. $H = (1.08)(6000)(150 - 60) = 583,200.$ Btu/hr (Equation 2-4)
4. $V_e = \dfrac{583,200}{(500)(20)} = 58.3$ gpm. (Equation 2-6)
5. $\Delta T_m = \dfrac{(180 - 60) - (200 - 150)}{2.3 \log (180 - 60)/(200 - 150)} = 80.1°\text{F}.$ (Equation 4-7)
6. Assume normal service, no corrosives; select 0.024-in.-thick wall copper tubing. Assume two-row coil to handle entire flow. Coils staged in series to achieve heat exchange, resulting in counter-flow conditions. Enter Figure 5.9-2; find conversion factor from volume flow (gpm) to velocity (fps) = 0.86.
7. Tube velocity $= \dfrac{(58.3)(0.86)}{(24)} = 2.09$ fps.
8. Enter Figure 5.9-3; using turbulators, find tube factor, $F_i = 63$.
9. Enter Figure 5.9-4; find fouling factor = 50.
10. Assume maximum heat transfer at closest fin spacing of 8 fins per in. or 96 fins per ft, add steps 8 and 9 for total factor of 113. Enter Figure 5.9-5c for 500 fpm velocity and find $U = 155$ Btu/hr/ft²/°F/row.
11. No. of rows $= \dfrac{583,200}{(12)(80.1)(155)} = 3.9$ or 4 rows.
12. Enter Figure 5.9-6; find $\Delta P_s = 0.058$ in. water/row.
    $\Delta P_s = (4)(0.058) = 0.23$ in. water at 70°F

178 INDUSTRIAL AND COMMERCIAL HEAT RECOVERY SYSTEMS

Nominal Or Ordering Width "B"	Nominal or Ordering Length "A"																		
	12"	18"	24"	30"	36"	42"	48"	54"	60"	66"	72"	78"	84"	90"	96"	102"	108"	114"	120"
12"	1.0	1.5	2.0	2.5	3.0	3.5	4.0	4.5	5.0	5.5	6.0	6.5	7.0	7.5	8.0	8.5	9.0	9.5	10.0
15"		1.88	2.5	3.13	3.75	4.33	5.0	5.63	6.25	6.88	7.5	8.13	8.75	9.38	10.0	10.63	11.25	11.88	12.5
18"		2.25	3.0	3.75	4.5	5.25	6.0	6.75	7.5	8.25	9.0	9.75	10.5	11.25	12.0	12.75	13.5	14.25	15.0
21"			3.5	4.38	5.25	6.13	7.0	7.88	8.75	9.63	10.5	11.38	12.25	13.13	14.0	14.88	15.75	16.63	17.5
24"			4.0	5.0	6.0	7.0	8.0	9.0	10.0	11.0	12.0	13.0	14.0	15.0	16.0	17.0	18.0	19.0	20.0
30"				6.25	7.5	8.75	10.0	11.25	12.5	13.75	15.0	16.25	17.5	18.75	20.0	21.25	22.5	23.75	25.0
33"					8.25	9.63	11.0	12.38	13.75	15.13	16.5	17.88	19.25	20.63	22.0	23.38	24.75	26.13	27.5

Figure 5.9-1. Finned-tube coil face areas. (*Courtesy The Trane Co., La Crosse, WI.*)

TUBE WALL THICKNESS	GPM CONVERSION FACTOR	
	1 ROW	2 ROW
0.024	1.72	0.86
0.035	1.85	0.93
0.049	2.05	1.03

$$\text{Tube Velocity} = \frac{(\text{gpm})(\text{conversion factor})}{\text{finned width}}$$

Figure 5.9-2. Finned-tube coil tube factors. (*Courtesy The Trane Co., La Crosse, WI.*)

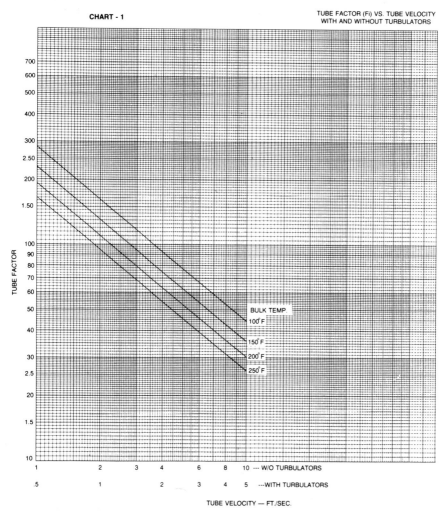

Figure 5.9-3. Finned-tube coil tube factor chart. (*Courtesy The Trane Co., La Crosse, WI.*)

180 INDUSTRIAL AND COMMERCIAL HEAT RECOVERY SYSTEMS

TEMPERATURE OF HEATING MEDIUM	UP TO 240° F		240° F TO 400° F	
TEMPERATURE OF WATER	125° F OR LESS		OVER 125° F	
WATER SOURCE	WATER VELOCITY FPS		WATER VELOCITY FPS	
	UP TO 3 FPS	OVER 3 FPS	UP TO 3 FPS	OVER 3 FPS
Sea water	50	50	100	100
Brackish water	200	100	300	200
Cooling tower				
Treated	100	100	200	200
Untreated	300	300	500	400
Hard (over 15 grains/gal)	300	300	500	500
Engine jacket	100	100	100	100
Distilled	50	50	50	50
Treated water feedwater	100	100	100	100

Figure 5.9-4. Finned-tube coil fouling factors and thermal resistances. (*Courtesy The Tubular Exchanger Manufacturers Assn., Tarrytown, NY.*)

13. Average $T_S = \dfrac{150 + 60}{2} = 105°\,F$.

14. $\Delta P_s = \dfrac{(0.23)(460 + 105)}{(530)} = 0.25$ in. water.

15. Enter Figure 5.9-7 at two-row, 24-in. header; find header pressure drop per coil section, by interpolation:

$$\Delta P_e = 2.41 + \frac{(5.49 - 2.41)(2.1 - 2.0)}{(3 - 2)} = 2.7 \text{ ft water.}$$

16. Enter Figure 5.9-8; find tube pressure drop per pass. There are two passes.

$$\Delta P_e = 1.236 + \frac{(2.651 - 1.236)(2.1 - 2.0)}{(3 - 2)} = 1.4 \text{ ft water/pass.}$$

17. Average temperature $= \dfrac{200 + 180}{2} = 190°\,F$

18. $\Delta P_e = \dfrac{(1.4)(530)}{(460 + 190)} = 1.1$ ft water per pass.

19. $\Delta P_e = (2)(2.7) + (2)(1.1) = 7.6$ ft. water.

20. Enter Figure 5.9-9 for coil dimensions.

Example 2: Direct contact gas/liquid heat exchanger, summer operation.

SIZING AND PERFORMANCE OF HEAT RECOVERY EQUIPMENT 181

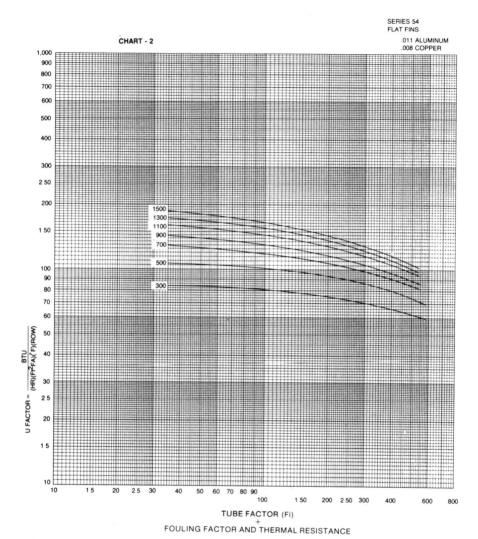

Figure 5.9-5a. Finned tube coil heat transfer coefficients. (*Courtesy The Trane Co., La Crosse, WI.*)

$$V_{se} = 55,000 \text{ cfm} \quad T_{se} = 95°\text{F}$$
$$w_{se} = 120 \text{ gr/lb}$$
$$V_{ee} = 50,000 \text{ cfm} \quad T_{ee} = 78°\text{F}$$
$$w_{ee} = 70 \text{ gr/lb}$$

182 INDUSTRIAL AND COMMERCIAL HEAT RECOVERY SYSTEMS

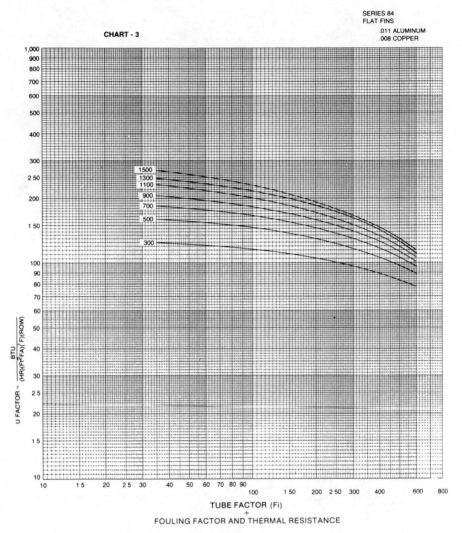

Figure 5.9-5b. Finned tube coil heat transfer coefficients. (*Courtesy The Trane Co., La Crosse, WI.*)

1. Air temperature difference $= T_{se} - T_{ee} = 95 - 78 = 17°\text{F}$.
2. Air humidity difference $= w_{se} - w_{ee} = 120 - 70 = 50 \text{ gr/lb}$.
3. Enter Figure 5.9-10; find sensible heat ratio $= 0.35$.
4. Enter Figure 5.9-11; find $E_e = 0.63$ or 63%, for smaller flow rate.
5. $E_s = \dfrac{(50,000)(0.63)}{(55,000)} = 0.57$ or 57%, for larger flow rate.

SIZING AND PERFORMANCE OF HEAT RECOVERY EQUIPMENT 183

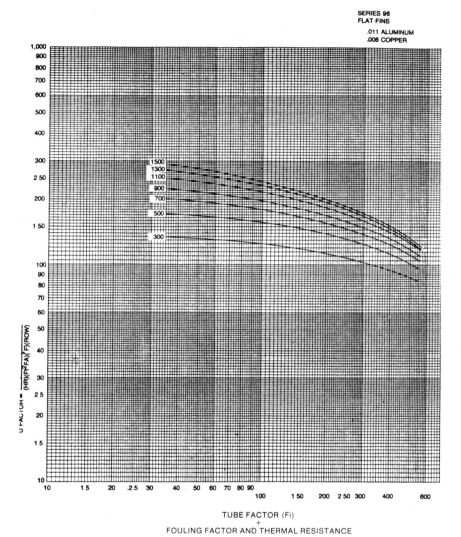

Figure 5.9-5c. Finned tube coil heat transfer coefficients. (*Courtesy The Trane Co., La Crosse, WI.*)

6. $T_{sl} = 95 - [(95 - 78)(0.57)] = 85°\text{F}.$ (Equation 4-8)
7. $w_{sl} = 120 - [(120 - 70)(0.57)] = 91.5 \text{ gr/lb}.$
8. Enter Figure 3.11-1; find $h_{se} = 42.2 \text{ Btu/lb}$; $h_{sl} = 35.2 \text{ Btu/lb}.$

Fin Type	Fin Series	300	400	500	600	700	800	900	1000	1200
Flat Fins	54	.013	.022	.031	.042	.057	.089	.092	.109	.149
	84	.022	.035	.050	.065	.088	.110	.135	.160	.215
	96	.025	.040	.058	.078	.100	.124	.150	.180	.240

Figure 5.9-6. Finned-tube coil air pressure drops. (*Courtesy The Trane Co., La Crosse, WI.*)

9. $H = \dfrac{(55{,}000)(0.076)(42.2 - 35.2)(60)}{(12{,}000)} = 146.3$ tons refrigeration (1 ton refrigeration = 12,000 Btu/hr).

10. Enter Figure 5.9-12; find supply unit = size 5000, face area = 120 ft².

11. $V_s = \dfrac{55{,}000}{120} = 458$ fpm.

12. Enter Figure 5.9-13 for vertical unit; find $\Delta P_s = 1.6$ in. water.

13. Repeat steps 6–12 for exhaust unit.

Example 3: Direct contact gas/liquid heat exchanger, winter operation with water added to solution.

Rows	Finned Width	Tube Velocity (FPS)							
		1	2	3	4	5	6	7	8
One Row	12	.27	1.15	2.56	4.51	7.15	10.05	13.85	18.02
	15	.22	.89	1.93	3.56	5.56	7.86	10.73	14.04
	18	.30	1.21	2.73	4.85	7.58	10.91	14.85	19.40
	21	.40	1.59	3.58	6.50	10.11	14.52	19.72	25.72
	24	.18	.75	1.65	2.91	4.52	6.61	8.94	11.63
	30	.12	.51	1.12	2.04	3.14	4.58	6.17	8.15
	33	.16	.58	1.34	2.33	3.69	5.25	7.21	9.33
Two Rows	12	.57	2.23	5.05	8.89	14.00	20.19	27.34	35.86
	15	.52	1.97	4.27	7.84	12.40	17.66	23.69	31.23
	18	.69	2.76	6.21	11.54	17.24	24.83	33.68	44.01
	21	.91	3.71	8.37	14.81	23.07	33.34	45.43	59.46
	24	.62	2.41	5.49	9.74	15.18	21.96	29.69	38.93
	30	.33	1.32	2.88	5.22	8.09	11.69	15.84	20.86
	33	.28	1.11	2.49	4.42	6.84	9.88	13.45	17.59

Figure 5.9-7. Finned-tube coil header pressure drops. (*Courtesy The Trane Co., La Crosse, WI.*)

Finned Length (in.)	Velocity With Turbulators (FPS)								Velocity Without Turbulators (FPS)							
	1	2	3	4	5	6	7	8	1	2	3	4	5	6	7	8
12	0.159	0.626	1.371	2.374	3.605	5.164	6.991	9.216	0.109	0.446	0.981	1.724	2.675	3.824	5.181	6.736
24	0.199	0.746	1.621	2.794	4.225	6.054	8.171	10.556	0.129	0.486	1.061	1.864	2.875	4.094	5.551	7.206
36	0.239	0.866	1.881	3.224	4.845	6.934	9.361	12.066	0.139	0.526	1.141	1.994	3.075	4.374	5.921	7.676
48	0.269	0.996	2.141	3.644	5.475	7.824	10.541	13.576	0.149	0.556	1.221	2.134	3.285	4.654	6.291	8.146
60	0.309	1.116	2.391	4.074	6.095	8.704	11.731	15.086	0.159	0.596	1.301	2.264	3.485	4.924	6.661	8.616
72	0.339	1.236	2.651	4.504	6.715	9.594	12.911	16.596	0.169	0.636	1.381	2.404	3.685	5.204	7.031	9.086
84	0.379	1.356	2.901	4.924	7.335	10.474	14.101	18.106	0.179	0.676	1.461	2.534	3.885	5.484	7.401	9.556
96	0.409	1.476	3.161	5.354	7.955	11.364	15.281	19.616	0.189	0.716	1.541	2.674	4.085	5.754	7.771	10.026
108	0.449	1.606	3.421	5.774	8.585	12.244	16.471	21.126	0.199	0.756	1.621	2.804	4.285	6.034	8.141	10.496
120	0.479	1.726	3.671	6.204	9.205	13.134	17.661	22.636	0.209	0.796	1.701	2.944	4.495	6.314	8.511	10.966
132	0.519	1.846	3.931	6.634	9.825	14.014	18.841	24.146	0.219	0.836	1.781	3.074	4.695	6.594	8.881	11.436
144	0.549	1.966	4.181	7.054	10.445	14.904	20.031	25.656	0.229	0.876	1.861	3.214	4.895	6.864	9.251	11.906

Figure 5.9-8. Finned tube coil tube pressure drops. (*Courtesy The Trane Co., La Crosse, WI.*)

186 INDUSTRIAL AND COMMERCIAL HEAT RECOVERY SYSTEMS

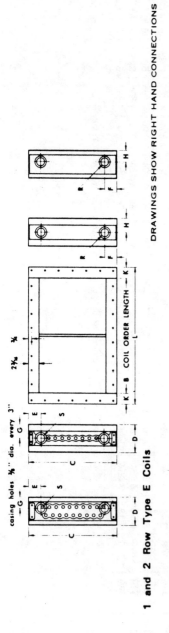

1 and 2 Row Type E Coils

DRAWINGS SHOW RIGHT HAND CONNECTIONS

Dimensions of Type E

A	C	D	\multicolumn{8}{c}{Type E 1 Row}	\multicolumn{8}{c}{Type E 2 Row}
			E	F	G	H	K	R	S	D	E	F	G	H	K	R	S
12	16½	6	2¾	2¾	3	3	2¼	1¼	1¼	6	2¾	2¾	2⅜	2⅜	2¼	1¼	1¼
15	19½	6	2¾	2¾	3	3	2¼	1½	1½	6	2¾	2¾	2⅜	2⅜	2¼	1½	1½
18	22½	6	2¾	2¾	3	3	2¼	1½	1½	6	2¾	2¾	2⅜	2⅜	2¼	1½	1½
21	25½	6	2¾	2¾	3	3	2¼	1½	1½	6	2¾	2¾	2⅜	2⅜	2¼	1½	1½
24	28½	6	2½	2½	3	3	2¼	2	2	6	2½	2½	2½	2½	2¼	2	2
30	34½	6	3¾	3¾	3	3	2¾	2½	2½	7¼	3¾	3¾	2¾	2¾	2¾	2½	2½
33	37½	6	3¾	3¾	3	3	3	2½	2½	7¼	3⅝	3⅝	2½	2½	3	3	3

For Type E coils, L = B + 2K

A = Coil order width   C = Overall casing width

Figure 5.9-9. Finned-tube coil dimensions. (*Courtesy The Trane Co., La Crosse, WI.*)

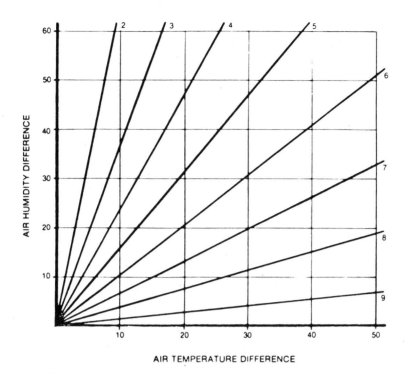

Figure 5.9-10. Direct contact gas/liquid heat exchanger sensible heat ratio. (*Courtesy Kathabar Systems, Midland Ross Corp., New Brunswick, NJ.*)

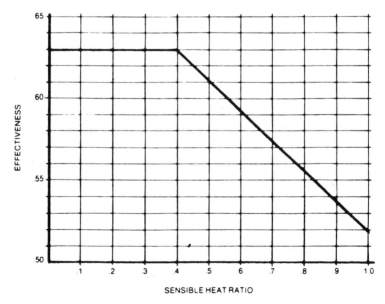

Figure 5.9-11. Direct contact gas/liquid heat exchanger effectiveness based on smaller flow rate. (*Courtesy Kathabar Systems, Midland Ross Corp., New Brunswick, NJ.*)

188 INDUSTRIAL AND COMMERCIAL HEAT RECOVERY SYSTEMS

*UNIT SIZE	AIRFLOW		WEIGHT				PUMP**			AIR FACE***
			HORIZONTAL T.C.		VERTICAL T.C.					
	MIN.	MAX.	SHIPPING	MAX. OPERATING	SHIPPING	MAX. OPERATING	H.P.	FLOW	PIPE SIZE	AREA (FT. SQ.)
2000	10,000	24,000	5,400	14,500	4,800	13,800	7.5	190	3	48
2500	12,500	30,000	6,200	17,000	5,600	16,400	10	240	3	60
3000	15,000	36,000	7,200	19,800	6,400	18,900	15	290	4	72
4000	20,000	48,000	9,000	25,100	7,900	23,800	15	385	4	96
5000	25,000	60,000	10,700	30,300	9,400	28,800	20	480	4	120
6000	30,000	72,000	12,400	35,500	10,800	33,700	20	580	6	144
7000	35,000	84,000	14,300	40,900	12,400	38,700	25	675	6	168
8000	40,000	96,000	15,900	46,000	13,800	43,600	25	780	6	192
9000	45,000	104,000	18,000	51,600	15,600	48,800	30	870	6	216

*Smaller unit sizes and non-standard sizes are also available.
**Based on 200 ft. equivalent run of pipe, two units at the same elevation and 70 ft. of pump head.
***Use air face area and actual airflow to determine unit pressure drop.

Figure. 5.9-12. Direct contact gas/liquid heat exchanger unit sizes. (*Courtesy Kathabar Systems, Midland Ross Corp., New Brunswick, NJ.*)

$$V_{se} = 55,000\,\text{cfm} \qquad T_{se} = 45°\,\text{F}$$
$$w_{se} = 10\,\text{gr/lb}$$
$$V_{ee} = 50,000\,\text{cfm} \qquad T_{ee} = 70°\,\text{F}$$
$$w_{ee} = 54\,\text{gr/lb}$$

Desired relative humidity of leaving supply flow = 80%, maximum permitted by unit.

1. Air temperature difference = $T_{ee} - T_{se} = 70 - 45 = 25°\,\text{F}$.
2. Air humidity difference = $w_{ee} - w_{se} = 54 - 10 = 44\,\text{gr/lb}$.
3. Enter Figure 5.9-10; find sensible heat ratio = 0.47.
4. Enter Figure 5.9-11; find $E_e = 0.61$ or 61%, for smaller flow rate.
5. $E_s = \dfrac{(50,000)(0.61)}{(55,000)} = 0.55$ or 55%, for larger flow rate.
6. Enter Figure 3.11-1; find $h_{ee} = 25.4\,\text{Btu/lb}$.
$h_{se} = 12.4\,\text{Btu/lb}$.
7. $h_{sl} = 12.4 + (0.56)(25.4 - 12.4) = 19.7\,\text{Btu/lb}$.
8. Enter Figure 3.11-1 at 80% relative humidity and 19.7 Btu/lb; find $T_{sl} = 52°\,\text{F}$, $w_{sl} = 46\,\text{gr/lb}$.
9. $H = (55,000)(0.076)(46.0 - 42.2)(60) = 953,040\,\text{Btu/hr}$.
10. $w_{ee} = 10 + (0.55)(54 - 10) = 34.2\,\text{gr/lb}$.

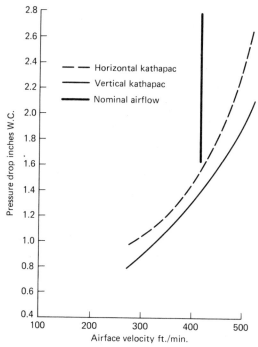

Figure 5.9-13. Direct contact gas/liquid heat exchanger pressure drops. (*Courtesy Kathabar Systems, Midland Ross Corp., New Brunswick, NJ.*)

11. Water taken up by supply flow = $\dfrac{(55,000)(46 - 10)(60)(0.076)}{(7000)} =$ 1289 lb/hr (7000 gr = 1 lb).

12. Water given up by exhaust flow = $\dfrac{(50,000)(54 - 34.2)(60)(0.076)}{(7000)} =$ 645 lb/hr.

13. Additional water required = $\dfrac{(1289 - 645)}{(60)(8.33)} = 1.3$ gpm.

5.10 BOILER ECONOMIZERS

Economizer on boiler stack
Boiler burns natural gas
Steam flow is 30,000 lb/hr
Blow-down is 1500 lb/hr

Gauge Pressure Lb./Sq. In.	Temp. °F	Volume Cu. Ft/Lb.	Total Heat Above 32°F/Lb.			Gauge Pressure Lb./Sq. In.	Temp. °F	Volume Cu. Ft/Lb.	Total Heat Above 32°F/Lb.		
			Sensible Heat h_f	Latent Heat h_{fg}	Total Heat h_g				Sensible Heat h_f	Latent Heat h_{fg}	Total Heat h_g
0	212.0	26.80	180.1	970.3	1150.4	200	387.8	2.14	361.8	837.5	1199.3
1	215.5	25.13	183.6	968.1	1151.7	205	389.8	2.09	364.0	835.7	1199.6
2	218.7	23.72	186.8	966.0	1152.8	210	391.6	2.04	366.0	833.9	1199.9
3	221.7	22.47	189.8	964.1	1153.9	215	393.6	2.00	368.0	832.1	1200.1
4	224.5	21.35	192.7	962.3	1155.0	220	395.4	1.96	370.0	830.4	1200.4
5	227.3	20.34	195.5	960.5	1156.1	225	397.2	1.92	372.0	828.6	1200.6
6	229.9	19.42	198.2	958.8	1157.0	230	399.0	1.88	374.0	826.9	1200.9
7	232.4	18.58	200.7	957.2	1157.9	235	400.8	1.85	375.9	825.2	1201.1
8	234.9	17.81	203.2	955.6	1158.8	240	402.6	1.81	377.7	823.4	1201.3
9	237.2	17.11	205.6	954.1	1159.7	245	404.4	1.78	379.6	821.9	1201.5
10	239.5	16.46	207.9	952.5	1160.4	250	406.0	1.74	381.5	820.2	1201.7
11	241.7	15.86	210.1	951.1	1161.2	255	407.7	1.72	383.3	818.6	1201.9
12	243.8	15.31	212.2	949.7	1161.9	260	409.3	1.69	385.1	817.0	1202.1
13	245.9	14.79	214.3	948.3	1162.6	265	410.9	1.66	386.9	815.4	1202.3
14	247.9	14.31	216.4	946.9	1163.3	270	412.5	1.62	388.6	813.8	1202.4
15	249.8	13.86	218.3	945.6	1163.9	275	414.1	1.60	390.4	812.2	1202.6
16	251.7	13.43	220.3	944.3	1164.6	280	415.8	1.57	392.1	810.6	1202.7
17	253.6	13.03	222.2	943.0	1165.2	285	417.2	1.54	393.7	809.1	1202.8
18	255.4	12.66	224.0	941.8	1165.8	290	418.8	1.52	395.4	807.6	1202.9
19	257.1	12.31	225.7	940.6	1166.3	295	420.3	1.50	397.1	806.1	1203.1
20	258.8	11.98	227.5	939.5	1167.0	300	421.7	1.47	398.7	804.6	1203.2
21	260.5	11.67	229.2	938.3	1167.5	305	423.1	1.45	400.3	803.1	1203.4
22	262.2	11.37	230.9	937.2	1168.1	310	424.6	1.43	401.9	801.6	1203.5
23	263.8	11.08	232.5	936.1	1168.6	315	426.2	1.41	403.5	800.1	1203.6
24	265.4	10.82	234.1	935.0	1169.1	320	427.5	1.39	405.0	798.6	1203.6

Temp	Press	v	h_f	h_fg	h_g
25	266.9	10.56	235.6	934.0	1169.6
30	274.1	9.45	243.3	928.9	1171.9
35	280.7	8.56	249.8	924.2	1174.0
40	286.8	7.82	256.0	919.8	1175.8
45	292.4	7.20	261.8	915.7	1177.5
50	297.7	6.68	267.2	911.8	1179.0
55	302.7	6.23	272.4	908.1	1180.5
60	307.3	5.83	277.2	904.6	1181.8
65	311.8	5.49	281.8	901.3	1183.1
70	316.4	5.18	286.2	898.0	1184.2
75	320.1	4.91	290.4	894.8	1185.2
80	323.9	4.66	294.4	891.9	1186.3
85	327.6	4.44	298.2	889.0	1187.2
90	331.2	4.24	301.9	886.1	1188.0
95	334.6	4.06	305.5	883.3	1188.8
100	337.9	3.89	308.9	880.7	1189.6
105	341.1	3.74	312.3	878.1	1190.4
110	344.2	3.59	315.5	875.5	1191.1
115	347.2	3.46	318.7	873.0	1191.7
120	350.1	3.34	321.7	870.7	1192.4
125	352.9	3.23	324.7	868.3	1193.0
130	355.6	3.12	327.6	865.9	1193.5
135	358.3	3.02	330.4	863.7	1194.1
140	360.9	2.93	333.1	861.5	1194.6
145	363.4	2.84	335.8	859.3	1195.1
150	365.9	2.76	338.4	857.2	1195.6
155	368.3	2.68	340.9	855.0	1195.9
160	370.6	2.61	343.4	853.0	1196.4
165	372.9	2.54	345.9	850.9	1196.8
170	375.2	2.47	348.3	848.9	1197.2
175	377.4	2.41	350.7	846.9	1197.6
180	379.5	2.35	353.0	845.0	1198.0
185	381.6	2.30	355.2	843.1	1198.3
190	383.7	2.24	357.4	841.2	1198.5
195	385.8	2.19	359.6	839.2	1198.8
325	428.9	1.37	406.6	797.2	1203.7
330	430.1	1.35	408.1	795.7	1203.8
335	431.6	1.33	409.6	794.3	1203.9
340	433.0	1.31	411.1	792.9	1204.0
345	434.3	1.29	412.6	791.5	1204.1
350	435.7	1.27	414.0	790.1	1204.1
355	437.0	1.26	415.5	788.7	1204.2
360	438.3	1.24	416.9	787.3	1204.2
365	439.5	1.22	418.4	785.9	1204.3
370	440.8	1.21	419.8	784.5	1204.3
375	442.0	1.19	421.2	783.2	1204.4
380	443.3	1.18	422.6	781.9	1204.4
385	444.6	1.16	423.9	780.6	1204.5
390	445.8	1.15	425.3	779.2	1204.5
395	447.1	1.13	426.7	777.3	1204.5
400	448.2	1.12	428.0	776.5	1204.5
405	449.4	1.11	429.3	775.3	1204.6
410	450.5	1.09	430.7	773.9	1204.6
415	451.8	1.08	432.0	772.6	1204.6
420	452.8	1.07	433.3	771.3	1204.6
425	453.9	1.06	434.5	770.1	1204.6
430	455.2	1.04	435.8	768.8	1204.6
435	456.3	1.03	437.1	767.5	1204.6
440	457.3	1.02	438.4	766.2	1204.6
445	458.4	1.01	439.6	765.0	1204.6
450	459.5	1.00	440.9	763.7	1204.6
455	460.6	.99	442.1	762.5	1204.6
460	461.7	.98	443.3	761.2	1204.5
465	462.7	.97	444.5	760.0	1204.5
470	463.8	.96	445.7	758.8	1204.5
475	464.9	.95	446.9	757.6	1204.5
480	466.0	.94	448.1	756.3	1204.5
485	467.1	.93	449.3	755.1	1204.4
490	468.0	.92	450.5	753.8	1204.4
495	469.0	.91	451.7	752.6	1204.3

Figure 5.10-1. Steam table. (*Courtesy Peabody Gordon-Piatt Inc., Winfield, KS.*)

192 INDUSTRIAL AND COMMERCIAL HEAT RECOVERY SYSTEMS

Temp. °F	Heat Content Btu/Lb	Temp. °F	Heat Content Btu/Lb	Temp. °F	Heat Content Btu/Lb	Temp. °F	Heat Content Btu/Lb	Temp. °F	Heat Content Btu/Lb
50	18.07	90	57.99	130	97.90	170	137.90	220	188.13
51	19.07	91	58.99	131	98.90	171	138.90	222	190.15
52	20.07	92	59.99	132	99.90	172	139.90	224	192.17
53	21.07	93	60.98	133	100.90	173	140.90	226	194.18
54	22.07	94	61.98	134	101.90	174	141.90	228	196.20
55	23.07	95	62.98	135	102.90	175	142.91	230	198.23
56	24.06	96	63.98	136	103.90	176	143.91	232	200.25
57	25.06	97	64.97	137	104.89	177	144.91	234	202.27
58	26.06	98	65.97	138	105.89	178	145.91	236	204.29
59	27.06	99	66.97	139	106.89	179	146.92	238	206.32
60	28.06	100	67.97	140	107.89	180	147.92	240	208.34
61	29.06	101	68.96	141	108.89	181	148.92	242	210.37
62	30.06	102	69.96	142	109.89	182	149.92	244	212.39
63	31.05	103	70.96	143	110.89	183	150.93	246	214.42
64	32.05	104	71.96	144	111.89	184	151.93	248	216.45
65	33.05	105	72.95	145	112.89	185	152.93	250	218.48
66	34.05	106	73.95	146	113.89	186	153.94	252	220.51
67	35.05	107	74.95	147	114.89	187	154.94	254	222.54
68	36.04	108	75.95	148	115.89	188	155.94	256	224.58
69	37.04	109	76.94	149	116.89	189	156.95	258	226.61
70	38.04	110	77.94	150	117.89	190	157.95	260	228.64
71	39.04	111	78.94	151	118.89	191	158.95	262	230.68
72	40.04	112	79.94	152	119.89	192	159.96	264	232.72
73	41.03	113	80.94	153	120.89	193	160.96	266	234.76
74	42.03	114	81.93	154	121.89	194	161.97	268	236.80
75	43.03	115	82.93	155	122.89	195	162.97	270	238.84
76	44.03	116	83.93	156	123.89	196	163.97	272	240.88
77	45.02	117	84.93	157	124.89	197	164.98	274	242.92
78	46.02	118	85.92	158	125.89	198	165.98	276	244.96
79	47.02	119	86.92	159	126.89	199	166.99	278	247.01
80	48.02	120	87.92	160	127.89	200	167.99	280	249.06
81	49.02	121	88.92	161	128.89	202	170.00	282	251.10
82	50.01	122	89.92	162	129.89	204	172.02	284	253.15
83	51.01	123	90.91	163	130.89	206	174.03	286	255.20
84	52.01	124	91.91	164	131.89	208	176.04	288	257.26
85	53.00	125	92.91	165	132.89	210	178.05	290	259.31
86	54.00	126	93.91	166	133.89	212	180.07	292	261.36
87	55.00	127	94.91	167	134.89	214	182.08	294	263.42
88	56.00	128	95.91	168	135.90	216	184.10	296	265.48
89	56.99	129	96.90	169	136.90	218	186.11	298	267.53

Figure 5.10-2. Feedwater table. (*Courtesy Peabody Gordon-Piatt Inc., Winfield, KS.*)

Combustion efficiency is 76%
Stack temperature is 600°F
Steam pressure is 100 psig
Condensate return temperature is 150°F
Feedwater temperature is 220°F
Excess air is 16%

SIZING AND PERFORMANCE OF HEAT RECOVERY EQUIPMENT 193

1. Enter steam table, Figure 5.10-1, at 100 psig; find:
   $h_{cos} = 1189.6$ Btu/lb (h_g).
2. Enter feedwater table, Figure 5.10-2 at 150°F; find:
   $h = 117.89$ Btu/lb.
3. Net boiler output = 1189.6 − 117.89 = 1071.71 Btu/lb.
4. Boiler output = (30,000)(1071.7) = 32,151,300 Btu/hr.
5. Boiler input = $\dfrac{(32,151,300)(100)}{(76)}$ = 42,304,342 Btu/hr.
6. Manufacturer's $E = 65\%$ for natural gas.
7. Enter Figure 5.10-3; find 0.90 lb exhaust gas/1000 Btu fired input.
8. Enter Figure 5.10-4; find specific heat flue gas = 0.273 Btu/lb.
9. $\Delta T = \dfrac{(65)(600 - 220)}{(100)} = 247°$ F. (Equation 4-8)
10. $H = \dfrac{(0.90)(42,304,342)(0.273)(247)}{(1000)} = 2,567,361$ Btu/hr. (Equation 2-2)

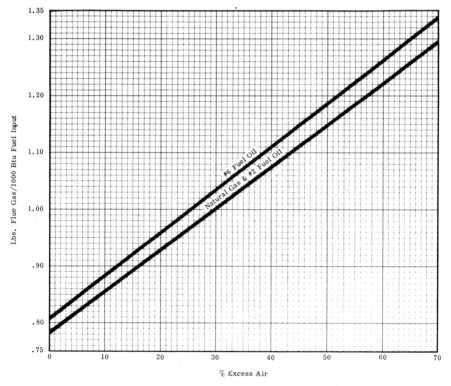

Figure 5.10-3. Pounds flue gas/1000 Btu fuel input. (*Courtesy Peabody Gordon-Piatt Inc., Winfield, KS.*)

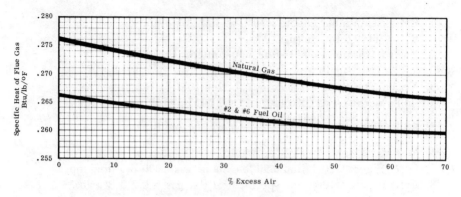

Figure 5.10-4. Flue gas specific heat. (*Courtesy Peabody Gordon-Piatt Inc., Winfield, KS.*)

11. $T_{el} = 600 - 247 = 353°F$.
12. $T_{se} = 220 + \dfrac{(2{,}567{,}361)}{(1)(30{,}000 + 1500)} = 302°F$. \hspace{1em} (Equation 2-2)

5.11 WASTE HEAT BOILER

Example: Multi-pass fire tube boiler

$$V_{ee} = 12{,}785 \text{ cfm} \qquad T_{ee} = 1600°F$$

Steam required at 150 psig.

1. $V_{\text{std},e} = (12{,}785)\dfrac{(530)}{(460 + 1600)} = 3{,}289$ standard cfm \hspace{1em} (Equation 2-5)
2. Assume for estimation purposes:

$$T_{ee} = 350°F \text{ for steam at low pressures.}$$
$$T_{ee} = 500°F \text{ for steam to 150 psig.}$$
$$T_{ee} = 550°F \text{ for steam above 150 psig.}$$

3. Specific heat at 1600°F estimated at 0.27 Btu/lb/°F.
4. $H = \dfrac{(1.08)(0.27)(3{,}289)(1600 - 500)}{(0.24)} = 4{,}395{,}748$ Btu/hr (Equation 2-4)

SIZING AND PERFORMANCE OF HEAT RECOVERY EQUIPMENT

$$= \frac{4,395,748}{33,500} = 131.2 \text{ boiler horsepower.}$$

33,500 Btu/hr = 1 boiler horsepower.

5. Enter steam table, Figure 5.10-1; find $h_{cos} = 857.2$ Btu/lb (h_{fg}).

$$\text{Steam evaporated} = \frac{4,395,748}{857.2} = 5128 \text{ lb/hr.}$$

6. Enter Figure 5.11-1; find 7.0 ft²/boiler horsepower.
   Boiler surface = (7.0)(131.2) = 918.4 ft².
7. V_e at 1600°F = 12,785 cfm. For other values of T_{ee}, find V_e, using Equation 2-5, at 1600°F.
8. Enter Figure 5.11-2, at model closest to required heating surface, HR-1000; find $\Delta P = 2.50$ in. water at 7,200 cfm.
9. $\Delta P_e = \left(\frac{12,785}{7200}\right)^2 (2.50) = 7.9$ in. water.

A boiler of this size will be 14 ft 5 in. long and 5 ft 9 in. in diameter, and will weigh 24,000 lb with water. If ΔP_e is excessive, a larger size can be chosen.

Heating Surface per Boiler Horsepower, ft²

PRESSURE RANGE P.S.I.	GAS TEMPERATURE, °F						
	2000	1900	1800	1700	1600	1500	1400
0–15	4.7	5.0	5.3	6.0	6.3	6.9	7.1
16–50	4.8	5.2	5.6	6.1	6.5	7.1	7.3
51–100	4.9	5.3	5.7	6.2	6.8	7.3	7.5
101–125	5.0	5.4	5.8	6.3	6.9	7.4	7.6
126–150	5.1	5.4	5.9	6.3	7.0	7.6	7.9
151–200	5.2	5.5	6.0	6.4	7.1	7.8	8.1
201–250	5.3	5.6	6.0	6.5	7.2	7.9	8.3

PRESSURE RANGE P.S.I.	GAS TEMPERATURE, °F					
	1300	1200	1100	1000	900	800
0–15	7.2	7.4	8.3	9.5	10.5	11.5
16–50	7.5	7.7	8.7	9.9	10.9	11.9
51–100	7.8	8.0	9.2	10.2	11.3	12.4
101–125	8.0	8.3	9.4	10.4	11.8	12.9
126–150	8.1	8.4	9.5	10.6	12.2	13.5
151–200	8.3	8.6	9.8	10.9	12.4	13.8
201–250	8.5	8.8	10.0	11.2	12.9	14.3

Figure 5.11-1. Multi-pass fire tube waste heat boiler heating surface per boiler horsepower. (*Courtesy York-Shipley, Inc., York, PA.*)

196 INDUSTRIAL AND COMMERCIAL HEAT RECOVERY SYSTEMS

MODEL	BASE CFM	BASE PRESSURE DROP
HR-125	900	.10″ W.C.
HR-150	1,080	.18″
HR-200	1,440	.38″
HR-250	1,800	.64″
HR-300	2,160	.53″
HR-350	2,520	.83″
HR-400	2,880	1.17″
HR-500	3,600	.60″
HR-625	4,500	.75″
HR-750	5,400	1.75″
HR-875	6,300	1.53″
HR-1000	7,200	2.50″
HR-1125	8,100	3.30″
HR-1250	9,000	1.40″
HR-1500	10,800	2.40″
HR-1750	12,600	4.00″
HR-2000	14,400	2.50″
HR-2500	18,000	2.50″
HR-3000	21,600	3.50″
HR-3500	25,200	4.00″
HR-4250	30,600	4.50″

Figure 5.11.2 Multi-pass fire tube waste heat boiler pressure drops. (*Courtesy York Shipley, Inc., York, PA.*)

5.12 LIQUID/LIQUID HEAT EXCHANGERS

Example:

$$H = 126,000 \text{ Btu/hr}$$

$$V_s = 25 \text{ gpm of water}$$

$$T_{se} = 70°\text{F}$$

$$T_{ee} = 150°\text{F} \qquad T_{el} = 135°\text{F}$$

$$P_e = 150 \text{ psig}$$

$$\text{Allowable } \Delta P_s = 5 \text{ psig}$$

$$\text{Allowable } \Delta P_e = 10 \text{ psig}$$

Supply water in tubes.
Exhaust water in shell.

MODEL IDENTIFICATION

TYPE CODE	SHELL DIAMETER CODE	SHELL LENGTH CODE	BAFFLE SPACING CODE	TUBE DIAMETER CODE	PASS CODE	OPTION CODE
F = fixed tube bundle with 150 psi 1040 kPa shell HF = fixed tube bundle with 250 psi 1725 kPa shell SSF = fixed tube bundle with 300 psi 2070 kPa shell and stainless steel materials	2 = 2.13 in 3 = 3.63 in 5 = 5.13 in 6 = 6.13 in 8 = 8.38 in 10 = 10.75 in	01 = 9 in 02 = 18 in 03 = 27 in 04 = 36 in 05 = 45 in 06 = 54 in 08 = 72 in 10 = 90 in	H = 1.13 in D = 2.25 in E = 4.50 in A = 9.00 in T = 15 in (approx.)	Y = 0.250 in OD R = 0.375 in OD C = 0.625 in OD	1P = one pass 2P = two pass 4P = four pass	CN = 90-10 copper nickel tube CNT = 90-10 copper nickel tube and tube sheet B = brass bonnet CNB = 90-10 copper nickel tube and brass bonnet CNTB = 90-10 copper nickel tube, tube sheet and brass bonnet

Figure 5.12-1. Shell-and-tube heat exchanger model identification. (*Courtesy Young Radiator Co., Racine, WI.*)

CAPACITY FACTOR and RATED PRESSURE LOSS of SHELL LIQUID

CAPACITY FACTOR Btu/min-F J/s-C @ liquid flow in shell, Fs, gpm l/s

RATED PRESSURE LOSS OF SHELL LIQUID psi kPa @ RATED FLOW gpm l/s

Normally Stocked Type (F / HF / SSF / CNT)	Base Model	Fs gpm	Fs l/s	CF Btu/min-F	CF J/s-C	Fs gpm	Fs l/s	CF Btu/min-F	CF J/s-C	Fs gpm	Fs l/s	CF Btu/min-F	CF J/s-C	Fs gpm	Fs l/s	CF Btu/min-F	CF J/s-C	Fs gpm	Fs l/s	CF Btu/min-F	CF J/s-C	psi	kPa	Rated Flow gpm	Rated Flow l/s
	Fs gpm l/s	2.0	0.13			4.0	0.25			6.0	0.38			8.0	0.50			10	0.63					6.0	0.38
– / • / • / –	201-HY			2.9	92			3.6	114			3.9	123			4.2	133			4.4	139	4.5	32		
– / • / •• / –	202-HY			5.8	183			7.2	228			7.9	250			8.3	262			8.7	275	9.1	63		
	Fs gpm l/s	3.0	0.19			6.0	0.38			9.0	0.57			12	0.76			15	0.95					9.0	0.57
– / • / •• / –	301-HY			6.3	199			7.7	243			8.7	275			9.4	297			9.9	313	2.6	18		
– / •• / ••• / –	302-HY			13	420			16	518			18	579			19	610			20	638	8.3	57		
– / – / – / •••	303-HY			20	635			25	777			27	863			30	935			31	980	14.2	98		
	Fs gpm l/s	6.0	0.38			12	0.76			18	1.1			24	1.5			30	1.9					18	1.1
– / – / – / –	301-DY			6.9	218			8.4	265			9.3	294			9.9	313			11	332	3.0	21		
– / • / •• / –	302-DY			14	439			17	531			19	588			20	629			21	657	6.0	42		
– / – / – / •••	303-DY			21	657			25	796			28	882			30	942			31	989	9.3	62		
	Fs gpm l/s	10	0.63			15	0.95			20	1.3			25	1.6			30	1.9					20	1.3
– / – / • / –	301-EY			6.4	202			7.3	231			8.0	253			8.5	269			8.9	281	0.5	4		
– / – / – / –	302-EY			13	408			15	465			16	506			17	547			18	566	1.1	8		
– / – / – / •••	303-EY			20	619			22	695			24	758			26	815			27	844	1.6	11		
	Fs gpm l/s	5.0	0.32			10	0.63			15	0.95			20	1.3			25	1.6					15	0.95
– / – / – / –	502-HY			33	1060			41	1310			46	1440			49	1540			51	1600	15.7	109		
– / – / •• / –	503-HY			51	1610			63	1990			69	2170			73	2300			76	2420	26.7	185		
– / – / – / •••	504-HY			69	2170			84	2670			93	2940			98	3090			104	3290	37.9	262		
	Fs gpm l/s	10	0.63			20	1.3			30	1.9			40	2.5			50	3.2					30	1.9
– / – / – / –	502-DY			35	1120			43	1340			47	1480			50	1570			52	1650	11.3	78		
– / • / • / –	503-DY			52	1650			64	2020			71	2230			75	2360			78	2450	17.1	118		
– / – / – / •••	504-DY			69	2190			85	2690			94	2970			100	3150			104	3290	22.8	157		
	Fs gpm l/s	15	0.95			25	1.6			35	2.2			45	2.8			55	3.5					35	2.2
– / – / – / –	502-EY			32	1000			38	1190			41	1290			44	1400			46	1460	2.2	15		
– / – / – / –	503-EY			48	1510			57	1790			61	1940			66	2070			69	2190	3.3	22		
– / – / ••• / •••	504-EY			64	2010			74	2340			82	2580			87	2760			92	2890	4.3	30		
	Fs gpm l/s	30	1.9			40	2.5			50	3.2			60	3.8			70	4.4					50	3.2
– / – / – / –	502-AY			32	1000			35	1100			38	1190			39	1240			41	1310	0.6	4		
– / – / – / –	503-AY			53	1670			57	1800			60	1910			63	2000			66	2090	2.0	13		
– / – / – / •••	504-AY			63	2000			70	2210			75	2360			78	2480			82	2600	1.2	8		
	Fs gpm l/s	5.0	0.32			10	0.63			15	0.95			20	1.3			25	1.6					15	0.95
– / – / – / –	602-HY			45	1410			56	1750			62	1960			67	2120			70	2220	8.8	61		
– / – / – / –	603-HY			70	2200			86	2710			97	3070			105	3290			110	3440	17.8	123		
– / – / – / –	604-HY			95	3000			115	3670			130	4110			140	4390			145	4610	27.2	188		
– / – / – / •	606-HY			140	4460			175	5590			195	6190			210	6570			—	—	40.6	280		

Fs gpm l/s												
602-DY	10	0.63	20	1.3	35	2.2	50	3.2	65	4.1	35	2.2
603-DY	45	1420	56	1760	65	2040	71	2260	75	2370	5.7	40
604-DY	70	2220	87	2750	100	3150	110	3410	115	3600	11.8	81
606-DY	94	2970	115	3670	135	4230	145	4580	155	4830	17.9	124
608-DY	145	4550	175	5560	205	6480	220	6950	230	7330	30.5	211
	190		235	7490	275	8630	300	9480			43.3	299
Fs gpm l/s												
602-EY	20	1.3	40	2.5	60	3.8	80	5.0	100	6.3	60	3.8
603-EY	50	1560	61	1930	67	2110	72	2260	75	2360	5.0	34
604-EY	74	2330	92	2910	100	3220	110	3440	110	3540	7.4	50
606-EY	99	3110	120	3820	135	4230	145	4520	150	4770	9.9	68
608-EY	150	4710	180	5750	200	6380	215	6760	225	7110	14.8	102
	196	6190	240	7650	270	8560	285	9010	305	9610	19.7	136
Fs gpm l/s												
602-AY	40	2.5	60	3.8	80	5.0	100	6.3	120	7.6	80	5.0
603-AY	49	1550	56	1760	61	1920	64	2020	68	2130	1.2	8
604-AY	81	2550	92	2900	97	3070	105	3250	110	3440	4.0	27
606-AY	98	3100	110	3540	120	3860	130	4040	135	4230	2.4	16
608-AY	150	4680	165	5280	185	5810	195	6160	200	6350	3.6	25
	195	6190	225	7080	240	7620	255	8090	270	8470	4.8	33
Fs gpm l/s												
301-HR	2.0	0.13	5.0	0.32	8.0	0.50	11	0.69	14	0.88	8.0	0.50
302-HR	3.7	117	5.0	158	5.8	183	6.3	199	6.6	207	1.6	11
303-HR	7.8	246	11	332	12	379	13	417	14	433	5.0	35
	12	373	16	515	19	585	20	626	21	664	8.6	60
Fs gpm l/s												
301-DR	10	0.63	15	0.95	20	1.3	25	1.6	30	1.9	20	1.3
302-DR	5.5	174	6.1	193	6.5	205	7.0	221	7.1	224	2.7	19
303-DR	11	348	12	386	13	414	14	436	14	452	5.5	38
	16	518	19	588	20	619	21	651	22	679	8.2	57
Fs gpm l/s												
301-ER	10	0.63	15	0.95	20	1.3	25	1.6	30	1.9	20	1.3
302-ER	4.5	142	5.0	158	5.5	174	5.8	183	6.2	196	0.4	3
303-ER	8.8	278	10	319	11	348	12	370	12	389	0.8	6
	13	417	15	474	16	518	16	553	18	578	1.2	8
Fs gpm l/s												
502-HR	4.0	0.25	8.0	0.50	12	0.76	16	1.0	20	1.3	12	0.76
503-HR	18	581	23	736	26	806	28	872	29	910	7.9	55
504-HR	28	894	35	1110	39	1230	42	1320	44	1390	13.4	93
	38	1210	47	1490	53	1670	56	1760	59	1850	19.1	132
Fs gpm l/s												
502-DR	8.0	0.50	18	1.1	28	1.8	38	2.4	48	3.0	28	1.8
503-DR	20	616	25	787	27	866	29	926	31	970	7.5	52
504-DR	29	913	37	1170	42	1320	44	1390	46	1460	11.3	78
	39	1220	490	1548	55	1740	59	1850	62	1940	15.1	104
Fs gpm l/s												
502-ER	15	0.95	25	1.6	35	2.2	45	2.8	55	3.5	35	2.2
503-ER	19	594	22	705	24	771	26	825	27	866	1.6	11
504-ER	29	904	33	1060	37	1150	39	1240	41	1300	2.5	17
	38	1190	44	1400	49	1550	53	1660	55	1720	3.2	22

Figure 5.12-2a. Shell-and-tube heat exchanger capacity factor and pressure loss. (*Courtesy Young Radiator Co., Racine, WI.*)

CAPACITY FACTOR and RATED PRESSURE LOSS of SHELL LIQUID

NORMALLY STOCKED TYPE			BASE MODEL	CAPACITY FACTOR Btu/min-F J/s-C @ liquid flow in shell, Fs, gpm l/s										RATED PRESSURE LOSS OF SHELL LIQUID psi kPa @ RATED FLOW gpm l/s			
F	HF	SSF*CNT**																
			Fs gpm l/s	25	1.6	35	2.2	45	2.8	55	3.5	65	4.1	45	2.8		
			502-AR	18	562	20	632	21	679	23	720	24	752	0.4	3		
			503-AR	29	926	33	1030	35	1120	37	1170	39	1220	1.2	8		
			504-AR	36	1130	40	1250	43	1360	46	1440	48	1530	2.2	15		
			Fs gpm l/s	5.0	0.32	10	0.63	15	0.95	20	1.3	25	1.6	15	0.95		
			602-HR	28	894	35	1120	39	1250	43	1350	45	1410	6.6	46		
			603-HR	44	1390	55	1740	60	1910	65	2060	68	2160	13.3	92		
			604-HR	59	1870	74	2340	82	2600	89	2800	93	2950	20.3	140		
			606-HR	89	2810	110	3510	125	3890	135	4200	140	4360	30.5	210		
			Fs gpm l/s	10	0.63	20	1.3	30	1.9	40	2.5	50	3.2	30	1.9		
			602-DR	28	885	35	1110	39	1240	42	1330	44	1400	3.3	23		
			603-DR	44	1380	54	1710	61	1910	65	2050	69	2170	6.6	46		
			604-DR	59	1870	73	2310	82	2600	89	2810	92	2910	10.2	70		
			606-DR	91	2890	110	3540	125	3920	135	4230	140	4420	17.3	120		
			608-DR	120	3820	150	4710	170	5340	180	5660	190	5970	24.6	170		
		• •	Fs gpm l/s	20	1.3	40	2.5	60	3.8	80	5.0	100	6.3	60	3.8		
			602-ER	31	986	39	1220	42	1340	46	1440	48	1510	3.7	26		
		•	603-ER	47	1490	57	1800	64	2010	68	2130	71	2250	5.5	39		
			604-ER	62	1970	77	2440	85	2690	92	2890	96	3020	7.4	51		
			606-ER	93	2940	115	3600	125	4010	135	4270	145	4520	11.1	76		
			608-ER	125	3950	155	4870	170	5340	180	5720	190	6070	14.7	102		
		• • • •	Fs gpm l/s	30	1.9	50	3.2	70	4.4	90	5.7	110	6.9	70	4.4		
			602-AR	29	900	33	1050	37	1170	39	1250	42	1310	0.7	5		
			603-AR	47	1480	54	1710	60	1900	64	2010	66	2100	2.3	16		
			604-AR	56	1770	66	2100	73	2310	79	2480	83	2610	1.4	9		
			606-AR	84	2660	100	3190	110	3480	120	3730	125	3920	2.1	15		
			608-AR	110	3540	135	4200	145	4610	160	5020	165	5210	2.8	20		
		• • •	Fs gpm l/s	15	0.95	30	1.9	45	2.8	60	3.8	75	4.7	45	2.8		
			802-DR	55	1720	68	2140	75	2390	81	2570	85	2690	4.6	32		
			803-DR	86	2730	105	3350	115	3670	125	3950	135	4200	9.4	65		
			804-DR	115	3670	145	4520	160	5020	170	5440	175	5560	14.3	99		
			805-DR	145	4580	180	5690	200	6260	215	6790	—	—	19.4	134		
			806-DR	175	5530	215	6790	240	7550	260	8180	—	—	24.5	169		
			808-DR	235	7430	290	9130	320	10100	—	—	—	—	34.8	240		

Fs gpm l/s													
802-ER	25	1.6	50	3.2	75	4.7	100	6.3	125	7.9	75	4.7
803-ER	57	1800	71	2230	78	2470	84	2640	88	2780	3.7	26
804-ER	90	2850	105	3320	120	3730	125	3950	135	4200	5.6	39
805-ER	115	3570	140	4460	160	4990	165	5250	175	5500	7.4	52
806-ER	140	4460	175	5560	195	6190	210	6670	220	6890	9.4	65
807-ER	170	5370	210	6600	235	7360	255	7990	265	8340	11.2	77
808-ER	225	7140	285	8940	315	9890	335	10500	350	11000	15.0	104
810-ER	285	8970	350	11000	390	12400	420	13200	—	—	18.8	130
Fs gpm l/s													
802-AR	50	3.2	90	5.7	130	8.2	170	10.7	210	13.2	130	8.2
803-AR	57	1790	68	2150	75	2370	81	2540	85	2670	1.4	10
804-AR	94	2970	110	3510	120	3860	130	4080	135	4230	4.8	34
805-AR	115	3570	135	4300	150	4770	160	5060	170	5310	2.9	20
806-AR	150	4800	180	5690	195	6190	210	6600	220	6920	6.0	41
807-AR	170	5400	205	6410	225	7170	240	7580	250	7960	4.3	30
808-AR	225	7140	270	8560	300	9450	325	10300	335	10600	5.8	40
810-AR	300	9510	360	11300	390	12400	420	13200	440	13800	12.1	83
Fs gpm l/s													
805-TR	60	3.8	100	6.3	140	8.8	180	11.3	220	13.9	140	8.8
806-TR	140	4390	165	5180	180	5690	195	6100	200	6380	2.2	15
807-TR	160	4990	185	5880	210	6570	225	7080	230	7330	1.6	11
808-TR	220	6950	260	8250	285	9070	305	9570	320	10100	2.9	20
810-TR	250	7960	305	9570	335	10500	360	11300	380	11900	4.3	29
Fs gpm l/s													
1005-AR	50	3.2	100	6.3	150	9.4	200	12.6	250	15.8	150	9.4
1006-AR	215	8630	265	8440	295	9350	325	10200	335	10600	6.8	47
1007-AR	250	7840	305	9610	345	10800	370	11800	390	12200	4.9	34
1008-AR	320	10000	410	12900	460	14600	490	15500	520	16500	6.4	44
1010-AR	440	13700	550	17200	600	18900	640	20300	680	21600	13.3	92
Fs gpm l/s													
1005-2R	150	9.5	200	12.6	250	15.8	300	18.9	350	22.1	250	15.8
1005-TR	245	7740	270	8530	285	8970	300	9390	310	9730	2.5	17
1006-TR	270	8470	290	9200	305	9640	320	10100	330	10500	5.5	38
1008-TR	305	9640	335	10600	360	11200	370	11600	390	12200	4.0	27
1009-TR	420	13300	460	14400	440	15300	510	16200	520	16500	7.3	50
1010-TR	500	15600	540	17000	570	18000	610	19200	620	19700	10.9	75
Fs gpm l/s													
1005-AC	50	3.2	100	6.3	150	9.4	200	12.6	250	15.8	150	9.4
1006-AC	110	3410	135	4300	150	4740	165	5180	170	5440	4.0	28
1007-AC	120	3860	155	4870	175	5500	190	5940	195	6230	2.8	19
1008-AC	165	5180	205	6480	230	7270	250	7930	265	8440	3.8	26
1010-AC	215	6790	270	8530	305	9640	325	10300	345	10900	7.8	54
Fs gpm l/s													
1005-TC	150	9.5	200	12.6	250	15.8	300	18.9	350	22.1	250	15.8
1006-TC	135	4300	150	4710	155	4900	165	5210	170	5370	3.2	22
1007-TC	155	4870	165	5280	180	5720	185	5910	195	6190	2.3	16
1008-TC	215	6760	230	7300	245	7740	260	8250	270	8470	4.2	29
1010-TC	250	7960	270	8560	290	9160	305	9610	320	10000	6.3	43

Figure 5.12-2b. Shell-and-tube heat exchanger capacity factor and pressure loss. (*Courtesy Young Radiator Co., Racine, WI.*)

201

1. $V_e = \dfrac{126{,}000}{(500)(150 - 135)}$ (Equation 2-6)

   = 16.8 gpm.

2. $\Delta T_{sl} = \dfrac{126{,}000}{(500)(16.8)}$ (Equation 2-6)

   = 15°F.

   $T_{sl} = 70 + 15 = 85°$.

3. $\Delta T_m = \dfrac{(150 - 70) - (135 - 85)}{2.3 \log (150 - 70)/(135 - 85)}$ (Equation 4-7)

   = 63.9°F.

4. Capacity factor = $\dfrac{126{,}000}{(60)(63.9)}$

   = 32.9 Btu/min − °F.

5. Select tube diameter to be 0.250 in.; tube diameter code is Y from Figure 5.12-1. Enter Figure 5.12-2a for flow rate in shell (F_s) V_e = 16.8 gpm, capacity factor = 32.9, and Y code designation. A model 502-DY has adequate capacity and ΔP_e close to that required. Referring to Figure 5.12-1, this exchanger has a 5.13-in. diameter, an 18-in.-long shell, 2.25 in. of baffle spacing, and 0.250-in. tubes. Choosing type F produces a fixed tube bundle, capable of the required ΔP_e, and a stock size.

SHELL DIAMETER CODE	MAXIMUM LIQUID FLOW IN SHELL		PASS CODE	LIQUID FLOW IN TUBES				PRESSURE LOSS AT MAX. PER 01 SHELL LENGTH CODE	
				Minimum		Maximum			
	gpm	l/s		gpm	l/s	gpm	l/s	psi	kPa
2	11	0.69	1P	3.5	0.22	28	1.8	1.2	8.3
3	34	2.1	1P	9.0	0.57	71	4.5	1.2	8.3
			2P	4.5	0.28	31	2.0	2.3	16
			4P	2.2	0.14	17	1.1	6.8	47
5	72	4.5	1P	20	1.3	120	7.6	0.6	4.1
			2P	10	0.63	72	4.5	2.1	15
			4P	5.0	0.32	31	2.0	4.1	28
6	120	7.6	1P	30	1.9	250	15	1.1	7.6
			2P	15	0.95	120	7.6	2.7	19
			4P	7.5	0.47	61	3.8	6.7	46
8	220	14	1P	57	3.6	460	29	0.6	4.1
			2P	29	1.8	180	11	1.1	7.6
			4P	14	0.88	115	7.2	4.1	28
10	590	37	1P	100	6.3	680	43	0.6	4.1
			2P	50	3.2	340	21	1.5	10
			4P	25	1.6	170	11	3.7	26

Figure 5.12-3. Shell-and-tube heat exchanger flow limits and pressure loss in tubes. (*Courtesy Young Radiator Co., Racine, WI.*)

SIZING AND PERFORMANCE OF HEAT RECOVERY EQUIPMENT 203

6. Correcting for design conditions:

$$\Delta P_e = \left(\frac{16.5}{30}\right)^2 (11.3)$$

$$= 3.4 \text{ psig.}$$

7. Enter Figure 5.12-3 for shell diameter code = 5, and V_e = 25 gpm. The minimum single-pass flow rate is 20 gpm and the maximum four-pass flow rate is 31 gpm. A compromise of a two-pass exchanger is chosen. The maximum two-pass flow rate is 72 gpm, and the ΔP per 01 shell length code is 2.1 psig.
8. Correcting for design conditions:

$$\Delta P_s = \left(\frac{25}{72}\right)^2 (2.1)(2)$$

$$= 0.50 \text{ psig.}$$

The shell length code in this example is 2, the last figure in the above calculation.

9. These calculations are repeated if other sizes or operating conditions are to be explored. This example is presented as an illustration of a preliminary estimation method. Manufacturers will supply computer-selected heat exchanger designs without charge.

5.13 AIR RE-CIRCULATION

Example 1: Ceiling fan in warehouse, roof area 100,000 ft². Ceiling temperature before fan installation = 90°F. Ceiling temperature after fan installation = 75°F. Roof construction insulated metal deck with built-up roof and metal girders, U = 0.16 Btu/ft²/°F/hr.
1. H = (0.16)(100,000)(90 − 75) = 240,000 Btu/hr. (Equation 4-5)

Example 2: Air returned from dust collector in place of makeup air.

$$V_e = V_s = 30,000 \text{ cfm}$$
$$T_s = 40°F \text{ (winter average in NJ)}$$
$$T_e = 75°F$$

1. H = (1.08)(30,000)(75 − 40) = 1,134,000 Btu/hr. (Equation 2-4)
2. Fuel saved at 80% heater combustion efficiency = $\frac{(1,134,000)(100)}{(80)}$ = 1,417,543 Btu/hr.

6
EXAMPLES OF HEAT RECOVERY SYSTEMS

Previous chapters have discussed the survey of a facility to locate sources and uses for heat recovery, the criteria for matching sources and uses, and the heat recovery equipment available for use with the matched source and use to form a complete heat recovery system. This chapter presents examples of heat recovery systems as a guide to assembling heat recovery systems. Some examples of systems that were not feasible are also presented. Using these examples, the reader will be able to put together his own system, and to calculate its performance.

6.1 METAL CASTING PLANT

A precision metal casting plant was surveyed for waste heat recovery. Two different types of sources were present. The first was a large, gas-fired mold heating furnace. The second was cooling water from four sources. Two separate projects evolved.

The stack temperature of the furnace was measured with a dial thermometer at the roof level and was found to be 340° F. The average velocity, found by a Velometer traverse, was 1400 fpm. When corrected for temperature, the flow velocity was 1792 fpm. The stack base was 20 in. by 20 in., for an area of 2.93 ft.^{2} The flow rate was calculated to be 5250 cfm or 3478 standard cfm.

The plant consisted of a large manufacturing area of 43,000 ft^2, mostly open and unobstructed, with a furnace area off the main floor. Building heat and cooling were provided by a large number of rooftop units. No process use of the heat was available. The only use for recovered heat was building heat. This usually provides a small annual cost saving unless a large amount of heat with a simple system is used. The requirement for a simple system precluded supplying the heat to the rooftop units for use in the existing supply. The cost of connecting all the rooftop units with insulated ducting or piping and multiple instrumentation was too expensive.

An alternative plan was evolved to provide in-plant heating with six overhead unit heaters over the aisles of the plant and along convenient walls, where the piping could be hung. The unit heater fans were controlled by their own thermostats. The unit heaters were connected in a single piping loop around the plant, with a bypass to each heater.

The piping system included two circulating pumps, one as a spare. The pumps were controlled by a thermostat on the stack to operate only when the furnace was operating. An expansion tank, water makeup valve, pressure relief valve, and flow control valves were also supplied.

A finned-tube coil was placed in a duct leading from the existing furnace stack to a new stack and fan on the roof. A manual damper in the existing stack was provided to shut off flow from the stack and divert it to the heat exchanger. Flow through the heat exchanger was controlled by a face-and-bypass damper before the coil. The damper was motor-operated and controlled by a thermostat in the water pipe leaving the coil. If the water temperature exceeded the set point, the damper bypassed part of the hot gases around the coil and out the exhaust stack. This method of control requires the fan to be capable of operating at full stack temperature. It was, therefore, equipped with a heat slinger and high-temperature grease in the bearings. The coil supplied a constant flow of water to the unit heaters. In case of circulating pump failure or low water, the damper automatically bypassed the hot gases. If the damper failed, the pressure-relief valve operated to vent steam generated in the coil. The coil was designed for 400°F operation, and so could not be damaged by furnace over-temperature. If this had been a problem, the manual damper in the stack could have been converted to motorized operation, controlled by a stack thermostat set at a temperature slightly below the limit of the coil operating temperature.

The operating conditions were as follows:

Finned-tube coil:
3478 standard cfm of exhaust gas at 340°F, cooled to 150°F, 675,000 Btu/hr recovered. 33.8 gpm of water at 140°F, heated to 180°F.

Construction used:
36 in. × 36 in. 10 row, 6 fins per in., air pressure drop 1 in. water, water pressure drop 15 ft water.

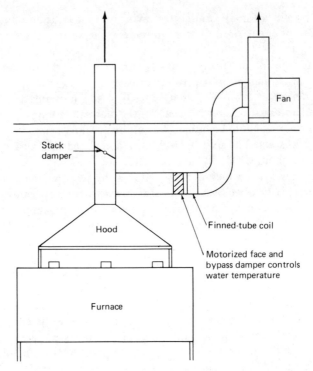

Figure 6.1-1. Finned-tube coil used with mold heating furnace.

Overhead heater:
Rated at 128,000 Btu/hr, 6 gpm water at 180°F, cooled to 140°F. Air flow rate of 3840 standard cfm at 60°F inlet, heated to 91°F outlet.

Fan:
5 hp, 20 in., backward inclined wheel, weatherproof construction and 400°F service. A drawing of the installation is shown in Figure 6.1-1. A subsequent addition to the system also provided sanitary water for the plant.

The same plant also had four sources of heated water: a 60 hp air compressor, an induction furnace, a vacuum melting furnace, and an air dryer. The total water flow rate was 63 gpm, heated to 90°F from 55°F. The water was entirely supplied from city mains and dumped into the sewer. The imposition of city sewer taxes and the increasing cost of city water indicated that some form of water re-circulation could produce considerable cost savings. As a source of cooling water, a well was available, producing brackish water. A system was designed using a water/water heat exchanger. The heat exchanger used the brackish well water to cool the system water. The system water was contained in a closed loop with two circulating pumps, an expansion tank and a make-up system.

The heat exchanger was constructed with a cast iron bonnet and cupronickel tubes. The shell was 8 in. in diameter, 90 in. long, and made from steel. The specifications were:

63 gpm of cooling water cooled from 90°F to 74°F.
69 gpm of brackish water heated from 58°F to 72°F.

6.2 ROOF TILE PLANT

The manufacture of asphalt roof tile or shingles involves the saturation of felt with hot asphalt. The saturated felt passes over steam-heated rollers during processing, and finally over chilled rollers prior to cutting to size. Heating the asphalt and saturating the felt produce objectionable hydrocarbon odors and heavy emissions, both of which must be eliminated because of air pollution regulations. One of several methods of air pollution control is the use of an incinerator.

An incinerator is essentially a large furnace. The hydrocarbon fumes are heated to the range of 1400°F to 1600°F for a short time. The hydrocarbons are burned to carbon dioxide and water vapor, which are odorless and invisible about the dew point. The incinerator is fired with gas or oil to raise the temperature of the gases to their ignition point and to provide heat not supplied by the burning hydrocarbon contaminants. The hot combustion products are exhausted to the atmosphere from the incinerator stack.

The steam for the rolls and for plant heat was supplied by a boiler supplemented by purchased steam. The steam was supplied at 150 psig through pressure reducing valves. In surveying the plant for heat recovery, replacing as much steam as possible through the use of a waste heat boiler produced the largest cost savings because of the large annual use of process steam. To size the boiler, an accurate measurement of the incinerator exhaust gas conditions was made.

Stack gas temperature measurements with a dial thermometer gave an average value of 750°F. Three traverses of the stack with a Velometer gave a measured value of 5000 fpm. When corrected for temperature, the corrected value was 8000 fpm. The stack size was 24 in. by 28 in., or an area of 4.67 ft.^{2} The resulting flow rate was 37,360 cfm or 16,360 standard cfm. The incinerator manufacturer estimated 14,000 standard cfm would be provided, or about 10% difference in values. For traverses under changing conditions and with large velocity variations from point-to-point in the survey, this is an acceptable difference.

The incinerator exhaust was rated by the manufacturer at less than 0.01 grains/ft^3, which would not cause an opacity problem. However, the sulfur content of the asphalt being heated was over 1%. When added to the sulfur in the fuel oil, the total was about 2%, or enough to cause sulfuric acid attack to the boiler tubes if condensation occurred in the waste heat boiler. The boiler exhaust temperature was kept well over 400°F.

The waste heat boiler selected for this application by the boiler manufacturer produced 5800 lb/hr of steam at 140 psig with a waste gas pressure drop of 3 in. water. It was supplied with the following equipment:

1. Shell 68 in. diameter with 30 in. diameter flanged inlet and 24 in. diameter flanged outlet, 6 in. flanged steam connection, 2-2-in. feedwater connection, 2 relief valves, 2-in.-thick fiberglass insulation with aluminum weatherproof jacket, removable full-diameter front and rear doors, manhole, and lifting lugs.
2. Water level control and low-water cutoff. The cutoff diverted waste gas flow from the boiler if the water supply failed. Water level gauge, try cocks, and gauge blow-down valves were also supplied.
3. Steam pressure gauge and control. Operated inlet and bypass valves by electric operators.
4. Blow-down valves removed dissolved solids.
5. High water level cutoff and alarm. Since the boiler was installed on an existing steam line, steam condensation could occur in the boiler when it was operating below existing steam temperature. The boiler would be blown down before the alarm system would allow re-start of the boiler.
6. Second low-water cutoff as a safety back-up for the boiler feedwater control.

The supply of boiler feedwater was provided by two turbine feed pumps operated from the boiler water level control. Because the water supply had a high alkalinity and iron content, a water treatment system was necessary. This consisted of a chemical treatment tank, additive pump, twin water softeners, and controls. The boiler was equipped with an automatic conductivity probe operating a continuous blow-down valve. The conductivity probe controlled the dissolved solids content of the boiler water.

The two stack method of control was used to connect the boiler to the incinerator. A sheet metal tee was connected to the incinerator blower discharge. The branch of the tee was connected to the boiler through an insulated duct, and the straight run of the tee formed a discharge stack. Electrically operated control valves in the stack and the branch were connected together and controlled by the boiler pressure. The valves were connected so that one opened as the other closed. The two control valves

EXAMPLES OF HEAT RECOVERY SYSTEMS 209

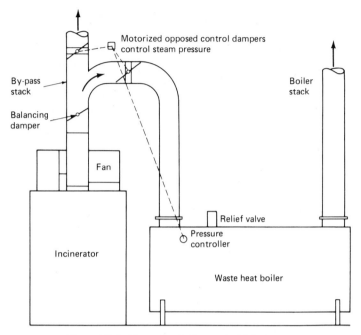

Figure 6.2-1. Waste heat boiler used with incinerator.

diverted only the amount of waste heat required to maintain steam pressure. In case of boiler outage, the boiler was closed off from the waste gases and the stack was opened for incinerator operation.

The boiler installation was unusual, in that it had to be placed on the roof of an adjacent building. The discharge of the incinerator was at roof level and ground space was unavailable. Additional bracing was required to support the filled boiler weight, and ground borings were required to verify load bearing qualities of the soil. Despite the complications of the installation, the large cost savings made it economically feasible, with the added advantage of increased cost saving as fuel prices increased. A drawing of the installation is shown in Figure 6.2-1.

6.3 MACHINE SHOP

A manufacturer of small metal parts faced the requirement of exhausting hot, contaminated air from his machine shop. Rather than pay for heated makeup air, a heat recovery system was installed to augment an existing makeup air system and allow it to handle the total load. To minimize the amount of air to be removed, two large hoods were placed in the hottest areas. These hoods

210 INDUSTRIAL AND COMMERCIAL HEAT RECOVERY SYSTEMS

were connected to a common duct, which went through the roof into an air/air heat exchanger. Following the heat exchanger, the cooled exhaust air was exhausted by a fan.

Incoming outdoor air was pulled through the heat exchanger by a fan and ducted to the inlet of the existing air makeup unit. An opening in the inlet duct with a damper after the heat exchanger was provided, so that the outdoor air could be brought into the existing makeup unit without going through the heat recovery system.

The system specifications were as follows:

Exhaust air—6000 cfm at 72°F.
Supply air—5000 cfm at 20°F.
Effectiveness—72%.
Supply air to makeup unit—57°F.
Heat recovered—200,000 Btu/hr.
Fans—5 hp, 18-in. backward inclined wheel.

Because it was on the roof, the entire installation was sprayed with a urethane foam insulation to reduce winter heat losses. The foam was $1\frac{1}{2}$ in. thick, and covered with a weather-proof mastic. Since the installation was used only in the heating season, the annual cost saving was not large. However, the fact that an additional heating unit was not required made the installation economically feasible.

6.4 TEXTILE DYE PLANT

A heat recovery system for a textile dye plant used hot water being discharged from several dyeing machines and a blow-down separator on the boilers. The hot water was used to rinse the dyed fabric, and then dumped to a drain. The water was discharged at 190°F, and flowed at the rate of 33 gpm from each of two machines and 12 gpm from the third machine.

The boilers consisted of two 350-hp boilers and one 250-hp boiler, producing 125 psig steam. During the winter all boilers operated 24 hours a day, but during the summer, only one of the larger boilers was operated with the second as a standby. The stack temperature was 400°F. Use of an economizer was considered, but the combination of low stack temperature and a difficult installation made the project uneconomical. As an alternative, heat exchangers and the use of a flash economizer were investigated.

The larger two dyeing machines were considered as one unit with a total flow rate of 66 gpm of waste water. The exchanger was designed to operate at 160°F inlet to allow for cooling between the machines and the exchanger. The exchanger was a tube-and-shell exchanger constructed with brass tubes and a cast iron bonnet. The exchanger received city water at 55°F, heating it to 125°F while cooling the waste water to 90°F. The heat recovered was 2,300,000 Btu/hr, and the installation produced an excellent cost saving.

The existing water heating system was modified with a thermostatic steam valve so that it would operate only to supply the additional temperature necessary following the heat exchanger on the recovery system. A circulating pump, water collection pit, and condensate collection system were also supplied.

The smaller machine was also supplied with an exchanger with the same temperature characteristics, to recover 400,000 Btu/hr, and the same auxiliary equipment. The cost savings were also excellent.

The flash economizer was sized to handle the two larger boilers on an average basis, that is one boiler full time and the second half time, for 18,000 lb/hr. In this case, the blow-down was estimated at 10% because of water conditions, or 1800 lb/hr of 125 psig water. The separator manufacturer quoted a unit to flash the water to 10 psig steam for the boiler de-aereator, saving 216 lb/hr of steam or 203,000 Btu/hr. The hot water heated 3 gpm of 55°F makeup water to 180°F, for an additional 200,000 Btu/hr saving. The additional equipment supplied with the flash economizer was: liquid level control, safety relief valve, low-level alarm, temperature control panel, and bypass for maintenance.

6.5 PET FOODS BAKING

A pet foods processor baked dog biscuits in an oil-fired oven. Following baking, the biscuits were dried in an oil-fired dryer. The dryer was directly heated and the oven was indirectly heated. The burners fired into a heating duct in the oven and exhausted through two stacks. The two stacks from the burners could be used for heat recovery, since the exhaust gas was clean and contained only combustion products. The oven vent stacks and the dryer vent stacks could not be used for heat recovery, because they contained a great deal of moisture and fats from the biscuits. Any attempt to cool the gases would have resulted in condensation of moisture and animal fat in the heat recovery system. The animal fat would turn rancid, become acidic, and quickly corrode the equipment. All stainless steel components would be required, making the cost unacceptable.

The dryer operated at 300°F, with supply air brought in through a filter above the conveyor belt entrance. The stack gas temperatures were high enough to pre-heat the dryer air to 300°F, but the dryer air inlet was 50 ft. and 120 ft., respectively, from the two oven stacks. A coil-loop system was selected, using three finned-tube coils. Two coils were used in the two combustion exhaust stacks and a third coil at the dryer inlet. Temperatures were measured at the two stacks with a dial thermometer, and Velometer traverses were made. Following correction, the exhaust flow data was:

Stack 1—932 standard cfm at 610°F.
Stack 2—854 standard cfm at 580°F.
Dryer—1974 standard cfm.

The two exhaust finned-tube coils were sized as follows:

No. rows—8.
Fin spacing—7 fins per in.
Size—18 in. by 18 in.
Air pressure drop—1 in. water.
Water pressure drop—10 ft water.
Exhaust gas cooled to 300°F.
Water flow rate—5 gpm.
Heat recovered—290,000 Btu/hr each stack, 580,000 Btu/hr total.
Water heated from 200°F to 325°F.

The dryer finned-tube coil was sized as follows:

No. rows—16.
Fin spacing—8 fins per in.
Size—24 in. by 24 in.
Air pressure drop—1 in. water.
Air heated from 40°F to 300°F.
Heat recovered—580,000 Btu/hr.
Water cooled from 325°F to 200°F.

Three fans were supplied, two for the hot stacks and one for the cold air inlet. The three coils were interconnected with a loop of 1-in. insulated pipe with an expansion tank, a pressure relief valve, and a circulating pump. Because of the high water temperature, a water-cooled packing box was specified for the pump.

The high water temperature required the system to operate at about 80-psig pressure. This is not an abnormal pressure and could be met with standard components. The alternative would have been to use a high-temperature oil. The necessity of careful assembly, joint leakage problems, and the cost of filling ruled out the use of oil. The total system had an excellent cost saving, as well as the added advantage that the pre-heated dryer air allowed higher dryer temperatures to be reached, with resulting increases in production and product quality. A piping schematic drawing is shown in Figure 6.5-1.

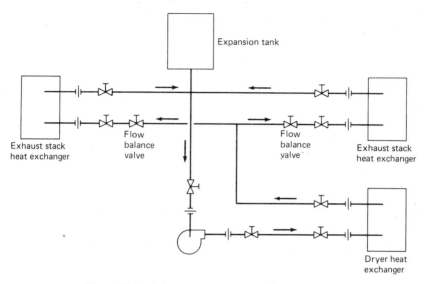

Figure 6.5-1. Piping schematic drawing for pet food dryer.

6.6 LARGE STEEL PLANT

The annealling furnaces of a steel mill were enclosed in a large, high-bay building. The furnaces exhausted directly into the building without exhaust stacks. As a result, the temperatures at the peak of the building were very high, measuring 105°F. At either side of the building were steam-heated wings where metal-forming operations were carried out. The steam heaters were overhead fan-coil units, each rated at 330 lb/hr of steam, or 305,000 Btu/hr. The units were 45 ft apart, at ceiling height.

A heat recovery system was designed to use the hot air from the peak of the furnace building to heat the wings. The hot air was brought from the peak area, through an insulated duct, over the roof, and down the side of the building to an air/air heat exchanger on the roof of the wing. The air/air heat exchanger heated outside air by cooling the hot building air. The heated outside air was brought through the wing roof and distributed at two locations from a ceiling duct.

The heat exchanger was sized as follows:

Flow rate—9000 cfm both flows.
Effectiveness—64%.
Air pressure drop—2.5 in. water on both flows.
Supply air heated from 0°F to 68°F.
Exhaust air cooled from 105°F to 37°F.
Heat recovered—664,020 Btu/hr.

Two steam heaters were eliminated. Ten-hp fans were provided for the exhaust and supply flows.

The furnaces ran continuously, providing recovered heat for the wings on a 24-hour basis during the winter. The system had the added feature that the installation could be expanded later to heat other areas of the wings with identical heat exchanger equipment.

In another section of the same plant, cast steel billets were ground to remove surface scale before heating for rolling. The grinding was performed under large hoods, which were exhausted to remove grinding dust. Each hood exhausted 30,000 cfm of heated air to a large dust collector. There were 20 hoods in the building.

A system was designed to bring 30,000 cfm of outdoor air through a fan on top of the hood, through the hood roof and into a long plenum the full width of hood at its face. The bottom of the plenum was open and equipped with movable guide vanes. By projecting the outdoor air down across the face of the hood, an air curtain was provided to accomplish the dust removal without using heated air.

Using 20° F average outside air temperature and 66° F room temperature, 1,500,000 Btu/hr were saved for each hood at 30,000 cfm.

This same method of using outdoor air for exhaust has been used successfully for laboratory hoods. The hood door can be kept closed and fumes still exhausted from the hood. It can also be applied to paint spray hoods. Outdoor air is brought through plenums along the top and sides of the hood. Slots and guide vanes project the air across the front of the hood to provide the required amount of exhaust air for the hood and avoid the exhaust of heated room air.

6.7 FORGING FURNACE

A plant heated forging blanks in an oil-fired pusher furnace. The furnace exhausted through flues into the forge room, producing a heat and smoke problem. At high fire, flames issued from the flues, greatly increasing heat and smoke. The forge room was heated with overhead unit heaters. An adjacent office building was heated by a hot water boiler. A heat recovery system using a collection hood over the furnace would both reduce the heat and smoke problems and capture wasted heat.

Before designing the system, the amount of heat available from the furnace had to be measured. Ideally, the temperature of the flue gases could be measured and the flow calculated from the oil consumption. The latter step proved impossible because the oil flow was changed with size of forging blank and operating temperature, and there was no furnace oil meter. It became necessary to make a traverse of the flues during high and low-fire conditions, then take an average for the amount of heat recovered. The recovery equipment, however, had to be designed for the high-fire conditions to avoid over-temperature problems. During low-fire conditions, the traverse was not difficult, but during high-fire conditions a traverse in a flame was necessary. At best this had to be done very quickly and only approximately with an average taken for the flow. When checked later with the heat recovery system, the data was reasonably accurate. The sum of the flow rates was 398 standard cfm and the average temperature was 1151° F. Since the finned-tube heat exchanger was limited to 450° F operation, air dilution was used to cool the air, resulting in 1130 standard cfm at 450° F.

The final design incorporated a large stainless steel hood over the furnace. At the top of the hood, two ducts were fitted, one vertical and one horizontal. The vertical duct went through the roof to the fan inlet. The horizontal duct contained a vertical finned-tube coil. After the coil, the duct turned upwards, went through the roof and to the blower inlet. Both ducts were equipped with dampers. By adjusting the damper on the finned-coil duct, the proper flow through the coil was established. By adjusting the damper on the vertical duct,

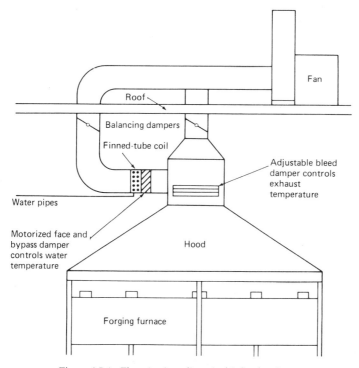

Figure 6.7-1. Finned-tube coil used with forging furnace.

the proper amount of diluting air was selected to keep the air temperature at high fire below 450° F. The water temperature was controlled by a motor-operated face-and-bypass damper in front of the coil. The damper position was controlled by a thermostat in the water line coming from the heat exchanger.

The hot water was piped to a storage tank, and then to the office building boiler and to the heaters in the forge shop. An aquastat on the tank controlled the circulating pump. Safety controls on the system included exhaust gas under-and-over-temperature, and water over-temperature. The exhaust gas under-temperature control prevented fan operation during periods of furnace shut-down to avoid cooling the tank water. The other two controls shut down the fan in case of water or exhaust gas over-temperture.

The finned-tube heat exchanger was sized as follows:

No. rows—8.
Fin spacing—6 fins per in.
Size—30 in. by 30 in.
Gas flow rate—1400 standard cfm at 450° F.
Water flow rate—13.4 gpm.
Water temperature rise—140° F to 190° F.

The fan was equipped with a heat slinger and high-temperature grease. A drawing of the installation is shown in Figure 6.7-1.

6.8 PLASTICS CURING OVEN

A plastic material was cured in a conveyor oven, operated at 215° F. The oven temperature was provided by heated air forced through a five-section steam coil by a fan. A second fan on the outlet end of the oven removed the hot air and chemicals released in the curing process, exhausting them to the outside. The supply air was pulled from the room at 5500 cfm. During warm weather, a welcome breeze came through the doors and windows. In the winter, the negative pressure condition created many cold drafts and forced the addition of an overhead steam heater in the room.

The heat recovery system installed on this plastics oven used a cross-flow air/air heat exchanger. The heat exchanger specifications were as follows:

Fin spacing—8 fins per in.
Effectiveness—61%.
Air flow rate—5500 cfm.
Exhaust temperature change—180° F to 100° F.
Supply temperature change—50° F to 130° F.

EXAMPLES OF HEAT RECOVERY SYSTEMS 217

Recovered heat—477,000 Btu/hr.
Pressure drop—2 in. water.

Supply air entered through a roof inlet, passed through the heat exchanger and into the supply fan inlet. From the supply fan discharge, the heated air entered the steam coils where the temperature was raised to the oven operating condition. The air leaving the oven passed through the exhaust fan, the air/air heat exchanger, and out the exhaust stack at a distance from the inlet stack and at a higher elevation to avoid re-circulation.

A second supply duct opening, inside the building, was controlled by a manual damper. During the summer, this damper was opened to evacuate hot air from the oven room. During the winter, it was closed to bring in outdoor supply air. The negative building pressure condition was entirely eliminated. The steam heating coils were equipped with a pneumatically operated temperature control, throttling the steam to supply only the additional required heat to reach operating temperature. The control also provided an accurate means of increasing oven temperature to experiment with new products and to increase production.

New fans were supplied to overcome the heat exchanger pressure drop and to replace the existing fans, which were in poor condition. The exhaust fan was in restricted quarters, resulting in overheating of the motor inside the dust

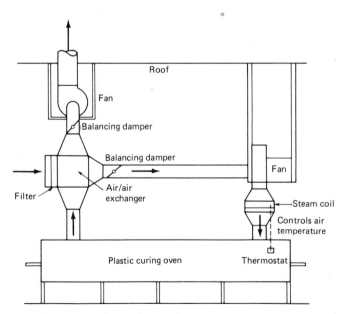

Figure 6.8-1. Cross-flow exchanger used with plastic curing oven.

218 INDUSTRIAL AND COMMERCIAL HEAT RECOVERY SYSTEMS

cover. To ease the temperature problem, ventilation slots were cut in the covers to provide better circulation.

Examination of the heat exchanger after a years' service showed no corrosion and very little deposition; they were cleaned in solvent and replaced in the housing. A drawing of the installation is shown in Figure 6.8-1.

6.9 BUILDING EXHAUST

A commercial building exhausted 39,000 cfm as part of its heating and ventilation system. The exhaust was a roof vent with an exhaust fan. The air makeup unit for the building was 40 ft away from the building exhaust on the same roof. Air was exhausted from the building continuously at 80°F. Night and weekend occupancy prevented any reduction in hvac system operation. The least expensive and least complicated installation used two coils and a connecting pipe loop. One coil was installed in the exhaust unit, and the other coil was installed in the air handler. The fans in each unit were speeded up to handle the additional pressure drop.

The coils were copper tube, aluminum finned, with the following specifications:

Width—90 in.
Length—126 in.
Rows—6.
Flow rate—39,000 cfm.
Pressure drop—0.6 in. water.
Water pressure drop—10.5 ft water.

The coil performance is as follows:

Winter supply:
Entering temperature—0.1°F.
Leaving temperature—38.5°F.

Winter exhaust:
Entering temperature—80°F dry bulb; 58°F wet bulb.
Leaving temperature—43°F dry bulb; 41.3°F wet bulb.
Heat recovered—1,624,400 Btu/hr.

Summer supply:
Entering temperature—95°F dry bulb; 76°F wet bulb.
Leaving temperature—87.1°F dry bulb; 73.9°F wet bulb.

Summer exhaust:
Entering temperature—80° F.
Leaving temperature—88.4° F.
Heat recovered—343,000 Btu/hr.

The circulating pump flow rate was 100 gpm. The circulating mixture was 30% glycol/water by volume to prevent freezing. The summer savings were balanced by the additional operating costs, but the winter savings were sufficient to make the system economically feasible.

The system was controlled by a three-way valve on the coil loop and a freeze-stat on the solution pipe to the exhaust coil. If the temperature of the exhaust coil solution dropped below the pre-set freezing limit, the three-way valve opened to bypass some of the circulating mixture away from the supply coil. The same valve can also be used to modulate the heat supplied during the winter, using a sensing bulb at the supply fan outlet and a pressure-selector control. A schematic drawing is shown in Figure 6.9-1.

The piping was run directly across the roof between the two air handlers. Double circulating pumps were supplied, one as a spare, and all pipes were insulated. The coils were mounted in the air handler housing, after some structural supports were added.

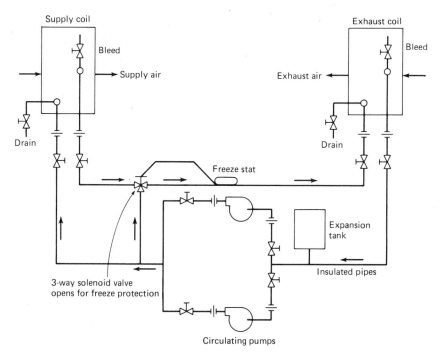

Figure 6.9-1. Coil loop piping for building heat recovery system.

6.10 HEAT RECOVERY SYSTEMS NOT FEASIBLE

Paint Dryer Oven

A heat recovery survey was made of large paint drying oven in an office furniture manufacturing plant. Heated air was supplied by a gas-fired air heater in the center of the oven roof. Exhausted heated air was to be picked up around the doors of the oven. The doors were slots in the oven walls to pass conveyorized parts. The application appeared to be favorable for heat recovery because the oven ran continuously, temperatures and flow rates were high, and no extensive piping or ducting was required. While the survey was in progress, an examination was made of the walls and exhaust areas of the ovens. A buildup of hardened paint or paint vehicle was observed, that could be removed only by cutting it off. The heat recovery equipment manufacturer had no previous experience in paint removal from his equipment, but was willing to accept the return of the equipment if the washing system did not work. Because the cost of installing the ducting system would not be repaid, and because of the equipment manufacturer's lack of direct experience, the system was not installed.

Chemical Factory

A chemical factory continuously exhausted air from laboratory hoods, and made up the air from a large air makeup unit. The laboratory exhaust and air makeup unit were within 30 ft of each other, making a coil-loop installation possible. The conditioned laboratory air was exhausted at 80° F both summer and winter. The flow rate was in excess of 20,000 cfm. Although the summer cost savings were not large, the winter cost savings were enough to make the heat recovery system economically feasible. There was concern for possible corrosion of the coil in the laboratory exhaust because of possible acid condensation from the exhausted laboratory fumes. A check of the humidity in the exhaust showed it to be very close to ambient humidity. The exhaust coil was sized to keep the exhaust air leaving the coil always above the dewpoint, even on wet, early spring days. A sample of the condensate was obtained by pulling an exhaust air sample through an ice-packed glass tube and collecting the condensate. It proved to be acidic, containing acid harmful to most metals. A stainless steel coil was out of the question economically. A sample of a baked phenolic coated coil fin was immersed for six months in the condensate and showed no attack. However, the question of acid attack still lingered and the heat recovery system was never installed.

Apartment House

A large boiler provided an apartment house with low-pressure heating steam. The apartment house had ample use for hot water in the hot water system. The burner was fired with #2 oil, making it a good candidate for heat recovery. The boiler flue temperature was checked at 340°F, which is very close to the limit for sulfuric acid corrosion. Because of this possible corrosion problem, no heat recovery system was proposed for the building.

Heat Treating Shop

A heat treating shop had many large gas furnaces used for hardening and annealling. These furnaces were both continuous and batch type, operating in a large, mostly unheated building. The heat-treating shop appeared to offer many opportunities for heat recovery because of the large number of furnaces and their high temperatures. As the survey progressed, the initial impression proved too optimistic. Because of the job shop nature of the shop, the furnaces were operated on differing schedules and at differing temperatures every day. The only use for recovered heat was for combustion air pre-heat, but the unsteady hours and temperatures of operation could not justify the expense of retrofitting the furnaces. Any collection of furnace exhausts into a single heat exchanger, and redistribution of the heated combustion air to the furnaces was prohibitively expensive and complicated. Therefore, the use of heat recovery had to be abandoned.

Large Bakery

A large bakery had several ovens that operated continuously and produced a clean, hot exhaust gas. Baked goods were stored in a large warehouse, heated by overhead, steam-fired unit heaters. The ovens were two floors and 600 ft from the warehouse. The heat recovery survey had to establish the warehouse heat loads, the oven stack exhaust heat, the cost of the extensive piping run between the oven and warehouse, and the cost of installing hot water heaters in the warehouse and a coil in the baking oven exhaust. Preliminary estimates were made and showed an excessive installation cost. Based upon these factors, the heat recovery system was abandoned.

6.11 SWIMMING POOL HEAT RECOVERY

Because of the high humidity levels generated in enclosed swimming pool structures, large amounts of outdoor air are circulated to maintain acceptable conditions. During cold weather, particularly, it is important to maintain close control of indoor relative humidity to avoid the annoyance and possible damage resulting from condensation on the inside surfaces of exterior walls and glass.

The use of a high percentage of outdoor air permits effective control of indoor relative humidity, but is an expensive proposition in cold climates

222 INDUSTRIAL AND COMMERCIAL HEAT RECOVERY SYSTEMS

unless some method is devised to recover heat from the exhaust or spill air and use it to help meet the space heating needs of the buildings. At a large New England community pool, a heat wheel proved to be the answer.

During occupied periods, up to 100 persons used the pool, and the pool was occupied 60% of the time. Pool water temperature was maintained at 78° F, primarily by the higher temperature air. A pool water heater provided supplemental heat if necessary. Ideally, the water temperature in a pool should be as low as possible, but comfortable, and the air temperature slightly higher to minimize evaporation from the pool surface. A high relative humidity, but a dewpoint low enough to prevent condensation in the pool area, is also desirable to minimize skin evaporation. The pool area was designed to be maintained at 80° F when occupied and 72° F when unoccupied. Because evaporation from the pool, splash water, and bathers acts as an evaporative cooler, heating was required during all but two or three weeks of the year.

To offset the moisture given off by evaporation in the pool, sufficient quantities of drier outside air were introduced, and a proportional amount exhausted after picking up moisture. The roof-mounted system is shown schematically in Figure 6.11-1.

During occupied periods, 7565 cfm of outside air was supplied to the pool building and a like amount exhausted, 6000 cfm from the pool area and 1565 cfm from separate locker and shower room exhaust fans. During unoccupied periods, a two-speed exhaust fan was switched to low speed, reducing pool exhaust to 3,000 cfm and makeup air supply proportionately to 4565 cfm. The fan switch is mounted on a wall in the pool area.

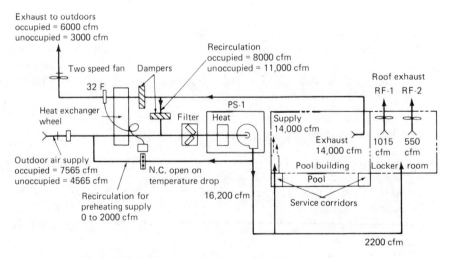

Figure 6.11-1. Schematic of swimming pool heat recovery system.

The heat wheel had a sensible effectiveness of 71%. Latent heat was not recovered. A re-circulation loop blended heated supply air with incoming outside air to protect against freezing. An automatic control modulated a damper to provide a safe mixture of heated and cold air. The amount of re-circulated air in the loop was less than 15% of the total supply air. Since the outdoor air is driest during freezing conditions, the drop in outdoor air supply did not affect condensation control in the pool.

An evaluation in the design stage revealed that heat recovery would result in substantial operating savings. Makeup air heating was required for all but three weeks of the year. On the other hand, the makeup air entering the heat wheel was pre-heated for less than three weeks of the year. A simple payback of 14 months was predicted, but the tendency of operating personnel to maintain a temperature almost 10°F higher than anticipated produced a much shorter payback period.

(*Information and illustration courtesy Heating/Piping/Air Conditioning magazine.*)

6.12 COMMERCIAL HEAT PUMP APPLICATIONS

Hospital Laundry

The laundry of a large medical center used more than 20,000 gallons of hot water a day. This had to be at 160°F so that when used in conjuction with detergents and bleach, it was hot enough to kill most germs or staph infections.

To reduce the amount of steam used to heat the water, the hospital installed a heat pump. The heat pump used available waste warm water at 90°F to 95°F to heat city water at 75°F to 150°F. Steam heating increased the water temperature to the 160°F needed by the laundry.

The daily cycle can be looked on as if the 10,000-gallon storage tank were drawn down twice each day. The tank was refilled with city water pre-heated to 75°F, and the tank contents gradually heated by circulating 62 gpm through the heat pump condenser. For the initial $2\frac{1}{2}$ hours after filling the tank, its contents were heated from 75°F to 105°F. Approximately seven hours were required to recover from 75°F to 150°F.

The output of the heat pump varied with the leaving condenser water temperature. The average output per hour and COP are properly determined by using the mean leaving condenser water temperature. At a mean temperature of 135°F, the heating capacity was 834,000 Btu/hr and the COP was 4.15. At an electricity cost of 4.86 cents/kilowatt hour and an oil cost of $5.72 per million Btu, the heat recovery system was expected to save $28,000 in its first full year of operation.

224 INDUSTRIAL AND COMMERCIAL HEAT RECOVERY SYSTEMS

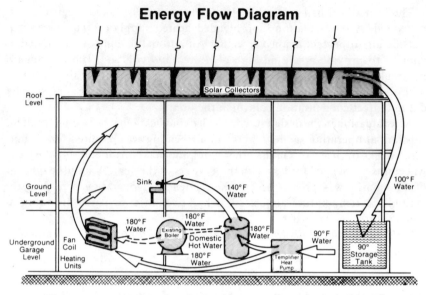

Figure 6.12-1. Schematic of typical heat pump-assisted solar hot water system.

Solar Heating

A heat pump has been used to assist a solar panel installation in Canada. It essentially doubled the solar system's capacity. The solar panel assembly consisted of 120 panels with a total area of 2300 ft.^{2} The panels were used in conjunction with a boiler to heat a block of 19 townhouses. In winter, the solar system provided 40% of the space heating and all of the hot water for the 19 units. In the summer, the excess heat was used to supply all the hot water for two adjacent blocks of 21 townhouses each. The heat pump had a capacity of 230,000 Btu/hr and a COP of 2.0. In this installation, the flat plate collectors operated at 70° F, approximately doubling the collector efficiency in January in Canada, over normal solar installations. The heat pump increased the water temperature to 180° F for space heating with fan coil units. A 7000-gallon storage tank in the underground garage was used to store up to two days of required hot water. A typical heat pump-assisted solar system is shown in Figure 6.12-1.

Hot water from the solar panels was stored in the storage tank at 90° F. The heat pump increased the water temperature to 180° F, and stored it in the domestic water tank, where it was blended to 140° F. Additional water at 180° F from the heat pump went to fan coil units for space heating. The boiler also supplied additional hot water to the fan coil units when the solar system capacity was exceeded.

(*Information and illustrations supplied by McQuay-Perfex Inc., Staunton, VA.*)

7
ECONOMICS OF HEAT RECOVERY SYSTEMS

Past chapters have been concerned with the selection, equipment choice, and performance calculations of a heat recovery system. The remaining steps are to find its cost, and determine if the cost is justified by the financial return. If several heat recovery systems are under consideration, the data developed in this chapter will also provide the means for ranking the systems according to their financial return.

7.1 ESTIMATION OF HEAT RECOVERY SYSTEM COST

Estimating heat recovery system cost can be handled in several ways. All methods start from a layout drawing of the proposed heat recovery system, a selection and description of the components, and the flows in the connecting pipes or ducts. The least complicated method of cost estimation is to turn it over to a consulting engineer. The consulting engineer will provide details of pipe and duct sizing, electrical services, and structural modifications, which can be sent out for bidding. The engineer can also provide his own cost estimate to be used in checking against bid prices. The consulting engineer, of course, will charge for his services, but in terms of ending up with a workable system, the money is usually well spent.

The next simplest method of system cost estimation is to call in a local contractor for a budget estimate. The contractor will use the system sketch, equipment description, and pipe or duct flows, to carry out the necessary pipe and duct sizing calculations, along with the building, structural, and electrical work estimates. His price estimate will be within normal budgetary accuracy of 10% to 20%, and there will usually be no charge for the cost estimates. Using a contractor as a cost estimator has a serious disadvantage. The budgetary estimate will probably contain no data on how the estimate was determined. Data on sizing, lengths, valving, etc. may be withheld by the contractor to protect his investment. If so, evaluating competitive bids will be very difficult, because there will be no basis for bid comparison. The budgetary estimate should be gone over with the contractor to insure that it is complete and workable.

The third method of cost estimation is to carry it out internally. Although more time-consuming, this method provides the best control over system costs by revising the system where required. Carrying out a system cost estimate requires a complete system design: mechanical, electrical, and structural. Piping and ducting is sized; electrical wiring runs and necessary equipment with electrical controls are planned; foundations, buildings and structures are designed.

For a simple heat recovery system without extensive structural and building modifications, the cost estimate can use data provided by standard estimating guides. These guides provide time and costs for a wide variety of ducting, piping, heating, electrical, and excavating jobs. One guide (Ref.1) breaks down the cost data into job units, crew size, material labor and overhead costs, and daily output. The system cost is estimated by listing all the work to be done in terms of the same job units as the estimating guide. The costs for material, labor, and overhead are listed from the guides as corrected for locality. The sum of all the costs plus a contingency is the estimated system cost. The contingency varies from 10% to 20%, depending on the completeness of the job listing. The sizing data for electricity wiring, water or liquid pipes, and gas or air ducting is available in engineering hand books and literature from suppliers. The heat recovery equipment manufacturers are a valuable source of help. They will often help make a system layout, and aid in the sizing of piping and ducting as part of their sales effort.

The division of responsibility for system design, construction and performance should be completely understood. The boundary of responsibility for system performance is especially important. Otherwise, future problems may involve the system designer, system installer, and equipment manufacturers at the same time, resulting in a passing of the responsibility on to the next party in a circular fashion.

7.2 PAYBACK

The evaluation of the estimated cost of the heat recovery system in relation to the cost savings from Equation 2-8 is carried out by calculating one or several financial ratios. These ratios are then compared with like ratios obtainable from other investment opportunities. If the heat recovery ratios are competitive with the same ratios from the other competitive investment opportunities, the heat recovery system clearly is feasible. If not, its pricing will have to be altered to improve its competitive position.

The least complicated and most widely financial ratio is the payback, as defined by Equation 7-1.

$$PB = \frac{I}{C - OE} \text{ years} \qquad (7\text{-}1)$$

The payback, PB, in years, is the investment, I, divided by the annual cost saving, C, less the operating expense, OE. The operating expense includes maintenance, operating costs for power, fuel, replaceable parts, etc. It does not include depreciation, taxes, investment credits, or increases in energy costs.

The payback is a very simple ratio, and serves as an initial screening method for proposed heat recovery systems. The payback is the time required to pay off the investment from the cost savings after subtracting operating costs. Acceptable values of payback range to four years, depending on the particular circumstances. Most acceptable values are in the two- to three-year range.

7.3 RETURN-ON-INVESTMENT

In its simplest form, the return-on-investment is the reciprocal of the payback (Equation 7-2).

$$ROI = \frac{(100)(C - OE)}{(I)} \text{ percent} \qquad (7\text{-}2)$$

The return-on-investment has the same use and limitations as the payback. Acceptable values are in the range of 30% or higher, depending on the circumstances. More complicated forms of return-on-investment are sometimes used, which will be explained later in this chapter.

7.4 CASH FLOW ANALYSIS

Both payback and return-on-investment are calculations based on a one-time investment and one-time cost saving. No consideration is given to the effect of time on the value of money, nor do they consider the effect of time on the value of the investment. Neither does either financial ratio take into account federal, state and local taxes, or special tax investment credits.

Energy Project Financial Evaluation Form

Date: _____ Project: _____ By: _____

Thousands of Dollars

	Item	0	1	2	3	4	5	6	7	8	9	10	Total
1	Savings @ ____ % escalation												
2	Maintenance & operating expense												
3	Operating income (1-2)												
4	Depreciation												
5	Taxable income (3-4)												
6	Income tax ____ (5)												
7	Net income (6-5)												
8	Cash flow in (7 + 4)												
9	Cash investment												
10	Start-up expense												
11	Investment tax credit .10(9)												
12	Energy tax credit ____ %(9)												
13	Salvage value												
14	Cash flow out (9 + 10–11–12–13)												
15	Net cash flow (14 + 8)												

Simple pay back = (9)/(11) _____ yrs

Net pay back = Time @ (15) = 0 _____ yrs

Return-on-investment = (100)(1)/(9) _____ %

Net return-on-investment = $(100)\dfrac{(7)}{10} / \dfrac{(14)}{2}$ _____ %

Average net return-on-investment = $(100) \dfrac{\text{Total }(7)}{\text{Life}} / \dfrac{(14)}{2}$ _____ %

Internal rate of return _____ %

Figure 7.4-1. Energy project financial evaluation form.

ECONOMICS OF HEAT RECOVERY SYSTEMS   229

All of these factors are incorporated into a calculation known as a cash flow analysis. This analysis is a time history of all the money that flows into or out of the heat recovery system during its lifetime. Using the data developed during the cash flow analysis, several financial ratios are calculated, which correct the deficiencies of payback and return-on-investment.

The method of calculating a cash flow analysis is illustrated by Figure 7.4-1. Each of the categories is discussed below. All calculations on Figure 7.4-1 are in real dollars.

Years

The number at the top of each column represents years since installation of the heat recovery system. The O column refers to the year of installation. The "total" column provides for the addition of yearly amounts. Figure 7.4-1 provides for up to ten years of system life. The number of years is best judged based on experience, or federal tax requirements. Ten years is a common number.

1. Savings at ___ *% Escalation.* Under column 1, the cost saving from Equation 2-8 is entered. In each successive year, this amount is increased by ___%. The ___% annual increase in the cost of energy is estimated based on local experience and the user's projections. (In 1981, ratios of 8% to 15% were in use.)

2. Maintenance and Operating Expense. Any expected maintenance charges, equipment servicing, parts replacement, electricity, gas, or oil costs, or operator costs are entered on line 2. These costs should be increased each year by an inflation rate, depending on economic forecasts.

3. Operating Income. In each year, line 2 is subtracted from line 1 to produce operating income.

4. Depreciation. The cash investment in the heat recovery system can be partially recovered from federal and state taxes on the income produced by the heat recovery system. Federal tax laws allow the investment to be depreciated by reducing the pre-tax savings. The depreciation is calculated in one of three ways. The choice of which to use depends on the method of accounting currently in use. If none is in use, the first is recommended as the simplest.

Straight Line

The cash investment in the heat recovery system is distributed evenly over the life of the system. Each year's depreciation is calculated by Equation 7-3

$$D = \frac{I}{L} \text{ dollars} \tag{7-3}$$

230 INDUSTRIAL AND COMMERCIAL HEAT RECOVERY SYSTEMS

The annual depreciation, D, is the investment, I, divided by the allowed equipment depreciation life, L. The depreciation is not increased by any inflation rate, since it is a fixed annual payment out of savings, and is put back into the cash flow. If there is any salvage value to the heat recovery system, its value is subtracted from the investment before solving Equation 7-3 for annual depreciation.

Sum-of-the-Years Digits

The straight line method of calculating depreciation does not take into account the increased depreciation experienced in the first years of the heat recovery equipment life. Tax laws allow two methods of calculating depreciation which more closely approximate the actual system depreciation. The first is called "Sum-of-the-Years Digits," and is based on Equation 7-4.

$$D = \left(\frac{L - Y + 1}{1 + 2 + \cdots + L}\right)(I) \text{ dollars} \qquad (7\text{-}4)$$

The year of calculation, Y, extends from 1 to the life, L. The numerator of Equation 7-4 represents the years of life remaining to the system in the year of calculation.

As an example the value of D for a $100 investment is calculated below:

YEAR	DEPRECIATION, DOLLARS
1	18.18
2	16.36
3	14.55
4	12.73
5	10.91
6	9.09
7	7.27
8	5.45
9	3.64
10	1.82
Total	100.00

Double-Declining Balance

The double-declining balance method of calculating depreciation produces the largest amount of depreciation of the three methods of calculation for the early years of system life. It is calculated by doubling the previous year's depreciated value, divided by the system life. The following example illustrates the method of calculation for a $100 investment.

YEAR	CALCULATION	DEPRECIATION DOLLARS
1	(100) (2)/10	20.00
2	(100 − 20) (2)/10	16.00
3	(80 − 16) (2)/10	12.80
4	(64 − 12.80) (2)/10	10.24
5	(51.20 − 10.24) (2)/10	8.19
6	(40.96 − 8.19) (2)/10	6.55
7	(32.77 − 6.55) (2)/10	5.24
8	(26.22 − 5.24) (2)/10	4.20
9	(20.98 − 4.20) (2)/10	3.36
10	(16.78 − 3.36) (2)/10	2.68
Total		89.26

The double-declining balance method does not produce a depreciation amount exactly equal to the cash investment. The remainder is assigned to salvage value.

5. *Taxable Income.* The taxable income is the operating income (3) less the depreciation (4).

6. *Income Tax.* The income tax is the taxable income (5) times the tax rate. The tax rate is the sum of applicable federal and state income taxes.

7. *Net Income.* The net income is the taxable income (5) minus the income tax (6). The annual net income for each year is summed and the total net income over the life of the heat recovery system is entered in the "total" column on line 7.

8. *Cash Flow In.* The cash flow returned to cash reserves is the sum of the depreciation (4) and the net income (7). The depreciation is an income because it is a repayment of the cash investment from the operating income (3).

9. *Cash Investment.* The cash investment is the estimated cost, and is entered in column O, enclosed in parentheses to indicate a negative number.

10. *Start-up Expenses.* If any one-time expenses are incurred because of system start-up, they are entered in column O and enclosed in parentheses as a negative number. These expenses include production interruption costs, special engineering costs, and start-up equipment.

11. *Investment Tax Credit.* A tax credit is allowed for capital investments of 10%. The amount entered in column O of line 11 is 10% of line 9.

12. *Energy Tax Credit.* Heat recovery systems (in 1981) are allowed an additional tax credit. The amount entered in column O of line 12 is the allowed percent of line 9, and varies according to the type of heat recovery equipment.

232 INDUSTRIAL AND COMMERCIAL HEAT RECOVERY SYSTEMS

13. Salvage Value. At the end of the system's life, a salvage value may exist for parts of the system. The salvage value will be made up of scrap metal and used equipment resale amounts. The salvage value is entered in column O of line 13.

14. Cash Flow Out. The cash flow out represents the cash flow out from cash reserves less any credits. It is equal to cash investments (9) plus start-up expenses (10) minus investment tax credit (11), energy tax credit (12) and salvage value (13). It is enclosed in parentheses to indicate a negative number.

15. Net Cash Flow. The net cash flow is the final summary of the cash flow both into and out of the cash reserves due to the heat recovery system, and is calculated for each year on line 15. The amount of line 14 is repeated in column O of line 15. The amount of cash flow on line 8 for each year, a positive number, is added algebraically to the value on line 15 for the previous year and entered on line 15. Negative values are enclosed in parentheses. At some point the value of line 15 usually becomes positive, and the parentheses are omitted. The values of net cash flow are used to calculate additional financial ratios which correct the omissions of payback and return-on-investment.

The data presented in lines 1 through 7 are a profit-and-loss statement, showing income from the heat recovery system. The data presented in lines 8 through 15 are a cash flow analysis, showing the annual cash flow resulting from the operation of the heat recovery system.

7.5 NET PAYBACK

The net payback refines the payback calculated in section 7.2 to include the data of Figure 7.4-1. It represents the payback time for the net cash flow, and is the time when the net cash flow becomes zero. The net payback is found from Equation 7-5.

$$NPB = Y + \frac{NCF_y}{CFI_{y+1}} \text{ years} \qquad (7\text{-}5)$$

The net payback, NPB, is the year, Y, of the last negative net cash flow from line 15, plus the fraction of that year's net cash flow, NCF_y, divided by the cash flow in the next year, CFI_{y+1}.

7.6 NET RETURN-ON-INVESTMENT

The net return-on-investment refines the return-on-investment calculated in Section 7-3 to include the data of Figure 7.4-1. It is calculated from Equation 7-6.

ECONOMICS OF HEAT RECOVERY SYSTEMS   233

$$NROI = \left(\frac{NI}{CFO}\right)(100) \text{ percent} \qquad (7\text{-}6)$$

The net return-on-investment is the value of net income, NI, in year 1 on line 7, divided by the value of cash flow out, CFO, in year 0 on line 14. Acceptable values of net return-on-investment are in the range of 30% or higher, depending on financial circumstances.

7.7 AVERAGE NET RETURN-ON-INVESTMENT

The average net return-on-investment refines the net return-on-investment calculated in Section 7.6 to include changes in net income and cash flow out over the system life. The average net income over the system's life is used instead of first year net income. The average cash flow out is used instead of first year cash flow out to reflect the effect of paying back the cash investment. It is calculated from Equation 7-7.

$$ANROI = \left(\frac{\text{Total } NI/L}{CFO/2}\right)(100) \text{ percent} \qquad (7\text{-}7)$$

The total net income (Total NI) in the "total" column of line 7, divided by the life L, forms the numerator of equation 7-7. The cash flow out (CFO) of line 14, column 0, divided by 2, forms the denominator. The result is the average net return-on-investment, $ANROI$.

7.8 DISCOUNTED CASH FLOW

The discounted cash flow calculation evaluates a heat recovery system as a revenue-producing investment. The calculation uses the annual cash flow in as the payments on the cash flow out. If a desired interest rate is specified, the calculation finds the amount of cash flow out paid for by the annual cash flow in payments over the heat recovery system's life. If a desired cash flow out is specified, the calculation finds the interest rate provided by the annual cash flow in payments such that the cash flow out is paid for over the heat recovery system's life.

The discounted cash flow calculation is a trial-and-error calculation, using a set of factors, called "Present Worth Factors" or "Present Value Factors."

$$PWF = \frac{1}{\left(1 + \dfrac{i}{100}\right)^Y} \qquad (7\text{-}8)$$

The interest rate, i, is called the "internal rate-of-return." The year of the calculation is Y. When PWF is multiplied by the annual cash flow in for the year, Y, of line 8 on Figure 7.4-1, the resulting dollar value is the value which, when invested at i interest, produced the annual cash flow in Y years later, and is called the "net present value." The factors for various values of i and Y are tabulated in many economic texts, and can be found in Ref. 2.

An example of the discounted cash flow calculation is given in Section 7.11. In the calculation, an internal rate-of-return is given or assumed. The annual cash flow in and PWF for each year of heat recovery system life are tabulated. Their products are tabulated in the next column and totaled at the bottom of the column. The sum is the net present value of all the annual cash flow in amounts at the given or assumed internal rate-of-return over the life of the heat recovery system. If the actual internal rate-of-return is to be found, several internal rate-of-return calculations are made at different values of i. The results are extrapolated to a zero value of net present value to find the actual internal rate-of-return.

Both payback and return-on-investment calculations are useful as internal financial ratios. They allow selection of heat recovery systems according to standards fixed by management or experience. Discounted cash flow calculations are useful as external financial ratios. They allow selection of heat recovery systems according to standards of maximum profits and interest rates, and allow comparison of the earned profit and interest rates with other investments. Each ratio has its advantages and disadvantages, and should be evaluated for a proposed heat recovery system.

7.9 NET PRESENT VALUE

The net present value of a heat recovery system must at least equal the cash flow out of the heat recovery system. This signifies to management that at an acceptable internal rate-of-return, i, the cash flow out investment is recovered. If the net present value is larger than the cash flow out, the excess represents a profit made on the heat recovery system. When several heat recovery systems are being considered, the system with the largest net present value should be chosen.

7.10 INTERNAL RATE-OF-RETURN

The internal rate-of-return of a heat recovery system must at least equal the interest charge paid for the cash flow out money. This signifies to management that the cost of borrowing money to pay for the heat recovery system is

ECONOMICS OF HEAT RECOVERY SYSTEMS 235

covered by the annual cash flow in payments. If the internal rate-of-return is larger than the prevailing borrowing rates, the excess rate is the profit of the heat recovery system investment expressed in percent interest. When several heat recovery systems are being considered, the system with the largest internal rate-of-return should be chosen.

7.11 EXAMPLE

Savings	$10,000 per year
Escalation	10%
Maintenance	$500 per year
Depreciation	Straight line
Income Tax	52%
Investment Credit	10%
Energy Credit	10%
Investment	$20,000

See Figure 7.11-1 for results.

Internal Rate of Return Calculation

YEAR	CFI	PWF @ 20%	(CFI)(PWF)	PWF @ 30%	(CFI)(PWF)
1	5.6	0.8333	4.66	0.7692	4.31
2	6.0	0.6944	4.17	0.5917	3.55
3	6.6	0.5787	3.82	0.4552	3.00
4	7.1	0.4823	3.42	0.3501	2.49
5	7.7	0.4019	3.09	0.2693	2.07
6	8.4	0.3349	2.81	0.2072	1.74
7	9.1	0.2791	2.54	0.1594	1.45
8	9.9	0.2326	2.30	0.1226	1.21
9	10.8	0.1938	2.10	0.0943	1.02
10	11.8	0.1615	1.91	0.0725	0.86
Total			30.82		21.70
			Net Present Value @ 20%		Net Present Value @ 30%

The actual internal rate of return is found where the *NPV* equals the *CFO*. This is found by interpolation

Energy Project Financial Evaluation Form

Date: _1981_ Project: _Heat Recovery System_ By: _SR_

Thousands of Dollars

	Item	0	1	2	3	4	5	6	7	8	9	10	Total
1	Savings @ _10_ % escalation		10.0	11.0	12.1	13.3	14.6	16.1	17.7	19.5	21.4	23.6	
2	Maintenance & operating expense		.5	.6	.6	.7	.7	.8	.9	1.0	1.1	1.2	
3	Operating income (1-2)		9.5	10.4	11.5	12.6	13.9	15.3	16.8	18.5	20.3	22.4	
4	Depreciation		2.0	2.0	2.0	2.0	2.0	2.0	2.0	2.0	2.0	2.0	
5	Taxable income (3-4)		7.5	8.4	9.5	10.6	11.9	13.3	14.8	16.5	18.3	20.4	
6	Income tax _.52_ (5)		3.9	4.4	4.9	5.5	6.2	6.9	7.7	8.6	9.5	10.6	
7	Net income (6-5)		3.6	4.0	4.6	5.1	5.7	6.4	7.1	7.9	8.8	9.8	
8	Cash flow in (7 + 4)		5.6	6.0	6.6	7.1	7.7	8.4	9.1	9.9	10.8	11.8	
9	Cash investment	(20)											
10	Start-up expense	0											
11	Investment tax credit .10(9)	2											
12	Energy tax credit _10_ %(9)	2											
13	Salvage value	0											
14	Cash flow out (9 + 10-11-12-13)	(16)											
15	Net cash flow (14 + 8)		(10.4)	(4.4)	2.2	9.3	17.0	25.4	34.5	44.4	55.2	67.0	63.0

Simple pay back = (9)/(1) __2.0__ yrs

Net pay back = Time @ (15) = 0 __2.67__ yrs

Return-on-investment = (100)(11)/(9) __50.0__ %

Net return-on-investment = $(100)\frac{(7)}{10}/\frac{(14)}{2}$ __22.5__ %

Average net return-on-investment = (100) Total $\frac{(7)}{Life}/\frac{(14)}{2}$ __78.8__ %

Internal rate of return __34.21__ %

Figure 7.11-1. Energy project financial evaluation form.

$$\frac{30.82 - 21.70}{30 - 20} = \frac{30.82 - 16}{x}$$

$(100) (20 + x) =$ internal rate-of-return $= 34.21\%.$

REFERENCES

1. Mossman, M. J. (Editor), *Mechanical and Electrical Cost Data*, R. S. Means Co. Kingston, MA. 1980.
2. Brown, R. J. and Yanuck, R. R., *Life Cycle Costing*, Fairmont Press, Atlanta, GA. 1980.

8
A HEAT RECOVERY SYSTEM CHECKLIST

A brief heat recovery system checklist is useful to insure covering the major areas of a heat recovery system design and installation.

Survey

Locate all sources and uses.
   List for each exhaust and supply:

   Composition.
   Flow rates.
   Temperatures.
   Contaminants.
   Operating hours
   Operating cycles.
   Size of pipe or duct connections.

   List for surroundings:

   Available structural supports.
   Access.
   Obstructions between sources and uses.
   Separation of source and use.

Matching

Compare survey of sources and uses to find best matching pair:

Least separation distance.
Most similar operating cycle.
Usable temperature range.
Reasonable use of recoverable heat.
Absence of contaminants, corrosives, condensibles.
Minimum installation problems.

Heat Recovery Equipment Selection

Locate suitable heat recovery equipment from Chapter 4. Calculate performance from Chapter 5 or with help of manufacturer.

Financial Evaluation

Layout heat recovery system.
Estimate system cost.
Calculate system cost savings.
Calculate financial ratios and make decision.

Installation

Make detailed drawings, purchase equipment, and install; or
Hire general contractor to carry out entire installation.
Avoid splitting responsibility.
Check performance of system at start-up.

INDEX

Absorption chemical apparatus, 31
Absorption refrigeration from recovered heat, 126
  cycle, 128
  sizes, 127, 130
  small unit, 130
Acidity, 17
Air pre-heat burner, 81
Air re-circulation, 112, 114
  ceiling fan, 112
  ceiling tube fan, 114
  floor mounted fan, 114
  performance calculation, 203
  vertical ducted ceiling fan, 113
Amount of heat, 5
Amount purchased, 12
Apartment house, 220
Approach temperature difference, 110
Average net return-on-investment, 233

Balance of heat, 34
Basic heat recovery, 1
Benefits of heat recovery, 3
Billed cost, 12
Blow-down economizer, 97
  performance, 99
Boiler economizers, 91
  blow-down economizer, 97
    performance, 99
  boiler stack economizer, 91
  construction, 92
  controls, 93
  performance calculation, 189
Boiler stack economizer, 91
  construction, 92
  controls, 93
  performance calculation, 189
Brake horsepower, 25
Building exhaust, 218
Burner with heat exchanger, 76
  performance calculation, 175

Calculation of cost saving, 11
Carbon dioxide volume, 28
Cash flow analysis, 227
Cash flow in, 231
Cash flow out, 232
Cash investment, 231
Ceiling fan, 112
  performance calculation, 203
Ceiling tube fan, 114
Centrifugal chillers and heat pumps, 119
  chiller, 119
    construction, 120
    cycle, 120
    storage, 122
  heat pump, 123
    COP, 126
    cycle, 123
    sizes and capacities, 126

Ceramic cross-flow heat exchanger, 67
  construction, 67
  performance, 67
  sizing calculation, 165
Change-of-state, 5
Checklist, 238
Chemical factory, 220
Chillers, 119
  construction, 122
  cycle, 120
  storage, 122
Coefficient-of-Performance (COP), 126
Combustion air pre-heat systems, 75
  air pre-heat burner, 81
  burner with heat exchanger, 76
  electronic controllers, 77
  elements, 75
  multiple burner installations, 76
  nozzle-mix burner, 81
  performance calculation, 173
  radiant tube furnace, 81
  self-recuperative burner, 78
  turndown, 81
Combustion exhaust flow rate, 28
Commercial heat pump applications, 223
Composition, 15
Condition of ducts, pipes, equipment, 16
Conduction, 36
Construction features, 17
Consulting engineer, 225
Contaminants, 16, 34
Contractor, 225
Convection, 37
Conversion factors, 11
Cross-flow heat exchanger, 38
Cross-flow tubular heat exchanger, 70
Counter-flow heat exchanger, 38
Counter-flow radiant tube heat exchanger, 68
  construction, 68
  performance, 69
    and sizing calculation, 167

Depreciation, 229
Dewpoint, 33
Dial thermometer, 20
Direct contact heat exchangers, 88, 90
  construction, 88, 91
  operation, 89
  performance and sizing calculation, 180
Discounted cash flow, 233

Division of responsibility, 226
Dry bulb temperature, 31
Double-declining balance depreciation, 230

Effectiveness, 40
Efficiency of heat source, 11
Electric heat, 12
Elements of a heat recovery system, 1
Emissivity, 37
Emitter, 37
Energy cost, 11
Energy prices, 11
Energy shortages, 4
Energy tax credit, 231
Engine and gas turbine heat recovery, 132
  gas turbine exhaust heat recovery, 134
    performance and size, 135
  heat recovery muffler, 134
  organic Rankine cycle, 135
    controls, 138
    operation and performance, 137
Enthalpy, 33
Escalation, 229
Estimating flow rate, 25
Estimating guides, 226
Estimation of heat recovery system cost, 225
Excess air, 28

Fan performance chart, 25
Film coefficient, 38
Finned-tube coil, 84
  circuits, 85
  coil-loop, 86
  construction, 84
  freezing, 86
  performance, 87
    and sizing calculation, 177
  size, 86
Fire-tube boilers, 99
  construction, 99, 101
  controls, 101
  multi-pass, 100
  performance and sizing calculation, 194
Floor mounted fan, 114
Flow, 21
Flow rates, 34
Flow volume rate, 22
Forging furnace, 214
Fundamentals of heat exchangers, 36

INDEX

Gas/gas cross-flow heat exchangers, 50
  construction, 51, 52
  performance, 51, 54
    and sizing calculation, 148
  size, 51
Gas/gas curved-plate counter-flow
    heat exchangers, 42
  construction, 43
  package, 45
  performance, 43
    and sizing calculation, 140
  size, 43
  waterwash system, 44
Gas/gas flat plate counter-flow heat
    exchangers, 46
  construction, 46, 49
  evaporative cooling, 49
  performance, 47
    and sizing calculation, 143
  size, 47
  waterwash system, 48
Gas/gas heat pipe heat exchangers, 57
  construction, 58
  package, 60
  performance, 59
    and sizing calculation, 156
  size, 59
  waterwash system, 60
Gas/gas high temperature heat exchangers, 66
  ceramic cross-flow, 67
    construction, 67
    performance, 67
      and sizing calculation, 165
  counter-flow radiant tube, 68
    construction, 68
    performance, 69
      and sizing calculation, 167, 168
  cross-flow tubular, 70
  stationary regenerative, 70
Gas/gas shell-and-tube heat exchanger, 54
  construction, 55, 57
  performance, 55, 57
    and sizing calculation, 153
  size, 55, 57
Gas/liquid heat exchangers, 84
  direct contact, 88, 90
    construction, 88, 91
    operation, 89
  finned-tube coil, 84
    circuits, 85
    coil-loop, 86
    construction, 84
    freezing, 86
    heat transfer fluids, 88
    performance, 87
    size, 86
  performance and sizing calculation, 177
Gas turbine exhaust heat recovery, 134
  size and performance, 135
Geographic features, 17
Greases, 17

Heat of vaporization, 7
Heat pipe boiler, 103
Heat pump, 123
  capacities and sizes, 126
  cycle, 123
Heat recovered by pre-heated combustion
    air, 7
Heat recovered from or by a gas, 6
Heat recovered from or by a liquid, 6
Heat recovered from or by a vapor, 7
Heat recovery measurements, 18
Heat recovery muffler, 134
Heat recovery survey, 13
Heat transfer coefficient, 37
Heat transfer fluids, 88
Heat treating shop, 221
Heat wheels, 61
  construction, 61, 62
  frosting, 66
  performance, 65
    and sizing calculations, 160
  purging, 62
  sealing, 62
  size, 61
Hospital laundry, 223
Hot water heat recovery units, 118

Income tax, 231
Installation, 35
Installation problems, 17
Internal rate-of-return, 234
Investment, 227
Investment tax credit, 231

Large bakery, 221
Large steel plant, 213
Latent heat, 33
Liquid flow, 21

Liquid/liquid heat exchangers, 107
  approach temperature difference, 110
  performance, 110
    and sizing calculation, 196
  piping, 111
  plate, 108
    construction, 109
  shell-and-tube, 107
    construction, 107
    multi-pass, 108
Location, 15
  of sources and uses of recovered heat, 13
Log mean differential temperature, 40

Maintenance expense, 229
Matching of source and use for recovered heat, 33
Metal casting plant, 204
Moist air conditions, 31
Moisture, 17
Multi-pass boilers, 100

Net cash flow, 232
Net income, 231
Net payback, 232
Net present value, 234
Net return-on-investment, 232
Nomenclature, ix
Nozzle-mix burner, 81

Operating cycle, 15, 33
Operating expense, 227
Operating hours, 11, 15
Operating income, 229
Organic Rankine cycle, 135
  controls, 138
  operation and performance, 137
Organic vapors, 17
Overall heat transfer coefficient, 38
Oxygen volume, 28

Paint dryer oven, 220
Parallel-flow heat exchanger, 38
Payback, 226
Percent saving of fuel, 7
Pet foods baking, 211
Pitot tube, 23

Plastics curing oven, 216
Plate liquid/liquid heat exchanger, 108
  construction, 109
Present value factor, 233
Present worth factor, 233
Pressure, 18
Pressure gauge, 19
Proximity, 34
Psychrometric chart, 31
Pump performance diagram, 25
Pyrometer, 20

Radiant tube furnace, 81
Radiation, 37
Receiver, 37
Recuperator, 41
Refrigeration heat recovery, 115
  hot water heat recovery unit, 118
  pressure-enthalpy diagram, 115
  schematic diagram, 117
  stainless steel double plate unit, 119
Regenerator, 41
Relative humidity, 31
Resistance thermometer, 20
Return-on-investment, 227
Roof tile plant, 207
Rotary regenerator, 41

Salvage value, 232
Saturation curve, 31
Self-curing compounds, 17
Self-recuperative burner, 78
  performance calculation, 176
Sensible heat, 33
Shell-and-tube liquid/liquid heat exchanger, 107
  construction, 107
  multi-pass, 108
  performance and sizing calculation, 196
Single-pass radiant tube heat exchanger, 69
  performance calculation, 173
Size of supply and exhaust systems, 15
Solar heating, 224
Solid particulates, 16
Source, 1
Sling psychrometer, 33
Specific heat, 5

INDEX

Specific volume, 33
Stainless steel double plate unit, 119
Standard conditions, 6
Standard density, 6
Standard specific heat, 6
Standard pressure, 6
Start-up expenses, 231
Stationery regenerator, 41, 70
Steam charts, 7, 191
Stefan-Boltzmann constant, 37
Straight line depreciation, 229
Subscripts, x
Sum-of-the-years digits depreciation, 230
Supply and exhaust operating conditions, 17
Surface temperature, 21
Swimming pool heat recovery, 221
System life, 229

Tachometer, 25
Taxable income, 231
Temperature, 5, 19
Temperature difference, 5
Temperature of source exhaust, 34
Textile dye plant, 210
Thermal conductivity, 37
Traverse, 22
Turndown, 81

Units, x
Use, 1
Use of this book, 4
U-tube manometer, 18

Velocity, 23
Velometer, 25
Vertical ducted ceiling fan, 113

Walking tour, 13
Waste heat boiler, 99
  ducting, 105
  fire-tube, 99
    construction, 99, 101
    controls, 101
    multi-pass, 100
  heat-pipe, 103
  performance and sizing calculation, 194
  water-tube, 102
    construction, 102
Water-tube boiler, 102
  construction, 102
Water vapor, 31
Weight flow, 5, 22
Well, 21
Wet bulb temperature, 31